Merry Christmas
2001
John Thornton Hilleary

White House Horticulture

White House Horticulture

MEMOIRS

by
John Thornton Hilleary

VANTAGE PRESS
New York / Washington / Atlanta
Los Angeles / Chicago

FIRST EDITION

Published by Vantage Press, Inc.
516 West 34th Street, New York, New York 10001

Manufactured in the United States of America
ISBN: 533-052834

Library of Congress Catalog Card No.: 81-90729

For my mother,
Mrs. Eleanor Trapnell Hilleary Dorsey Dosh

Contents

Acknowledgments

Foremost, I would like to thank my former White House supervisors, Bill Ruback and Irv Williams, who were instrumental in having me transferred to the formal White House gardens, without which occurrence this book would not have been written. Thanks go also to James F. Williams of the U.S. Geological Survey, who encouraged me to talk about my White House years on the horticultural staff and subsequently transform notes into a coherent document.

Introduction

Scarcely had I believed in the closing decade of the 1950s, when I was being taught the basics of gardening on my uncle's estate in North Baltimore, that I would one day reach the ultimate pinnacle in horticulture and landscape gardening. It was almost beyond comprehension when I was notified that I had been selected to assume the responsibility of the Presidential Gardens on the immaculate White House grounds in downtown Washington. Prior to my entrance onto the executive grounds, I had taken a job with the National Capital Parks and received additional training beyond that given me in Baltimore. I had spent the spring, summer, and a portion of the autumn season of 1965 assisting with foundation plantings around various government buildings in the Federal City. At the end of that year and having decided on a minimum of a two-year college degree in horticulture, I was transferred to the Kenilworth Aquatic Garden complex, where I would gain the necessary experience in preparation for a career in plant science. My days were numbered at Kenilworth, however, as my destiny lay eventually with the White House. Nevertheless, three one-year plant and tree propagation/planting cycles at Kenilworth hot house had given me the dexterity and knowhow to competently handle the floral assemblages growing in the Presidential Park, as it was referred to in the early 1970s.

I found the prestigious Rose and Jacqueline Kennedy Gardens of the White House quite to my taste and although the tempestuous pace of White House living may have brought on pangs of loathing, there was a certain defined measure of status attached to this coveted position. Status may have been reason enough to endure the everyday arduous rhythm. The proper touch, albeit watering, fertilizing and general pampering of the plants of the First Family, was entrusted to us; care of the elite and elegant bonsai trees given the President of the United States from the government of Japan; gentle and professional interior floral decoration of the White House for holidays and state commencements . . . and myriad numbers and varieties of Holland-grown tulips to grace the White House gardens were part of our horticulture menu.

This volume concerns the major species of flowers, shrubs and trees of the White House grounds, as seen through the eyes of a former White House gardener; plant maladies and treatment, tubculture in connection with portable decoration in and around the Executive Mansion, with some emphasis on the White House flower shop and herbculture.

Preface: A Brief History of the White House Gardens and Grounds

The earliest recollection of a garden at the White House dates to the turn of the eighteenth century, during the administration of John Adams. In those days an attempt was made to create and orient a thing of beauty amid the flotsam and jetsam of construction workers' sheds, brick kilns and general chaos strewed across what was to become the manicured South Lawn of later epochs.

As the new century marched forward, some attention was paid to the aesthetics of White House landscape and horticulture. The obnoxious tidal flats of the Tiber River Drainage Basin were filled in during the first quarter of the nineteenth century; the flats were located on parts of what today is the Lower South Grounds of the White House. President Jefferson was responsible for building the earth mounds south of the South Portico in the 1820s. This addition gave the South Grounds a gentle rolling park effect. The Civil War brought on not only misery for North and South alike, but also for the grounds of the White House. Union campfires singed the grass; horses' hoofs and the rutting effect of cannon wheels tore up the lawn on the south side of the mansion in a vigil determined to protect President Lincoln from the ravages of an indigenous war.

The third quarter of the nineteenth century marshalled in the construction of greenhouses and conservatories on the west end of the White House, where the West Wing stands today. These additions lasted until the first decade of the twentieth century.

Roses were planted on the White House grounds as early as 1913 by Ellen Axson Wilson, and this arrangement remained largely unchanged until the Kennedy administration, at which time the Rose Garden (West Garden, President's Garden) was redesigned by Mrs. Paul Mellon. The Rose Garden is typically eighteenth century American with traditional plantings of crab apples, holly osmanthus and English Boxwood.

Visitors to the White House are welcomed in the Rose Garden by the President of the United States. It is used for various delegations, press conferences, bill signings, and medals-of-honor ceremonies.

Directly to the east of the Rose Garden is the Jacqueline Kennedy Garden, so named by Mrs. Lyndon B. Johnson, and . . . also known as the East Garden or the First Lady's Garden, where the First Lady receives her guests. The Jacqueline Kennedy Garden is similar in style to the Rose Garden and boasts of extended rows of boxwood (green pillow species), holly osmanthus hedges, *Ilex convexa* (holly) and topiary holly overseeing square flower beds containing a profusion of spring bulbs in season; summer flowers—geraniums, daisies, zinnias and dusty miller. Autumn displays include chrysanthemums and heliotrope. The winter season brings color in the form of silver-gray St. Helenia herb planted in the Rose Garden rectangle forms under the flowering crab apples. Sleeping pansies occasionally awaken on a warm winter day with a colorful promise of springtime.

A portion of the Lower South Grounds of the White House is graced by the Children's Garden, a Christmas gift from President and Mrs. Johnson, given in 1968 in the memory of the nation's youngsters. This small garden is hidden among various evergreen species and, like the more formal gardens near the mansion, is planted with a seasonal variety of both annual and perennial flowers.

White House Horticulture

History of the Tulip Bulb

As the White House tourist gazes lovingly at the spring display of tulips from the Netherlands, few really know what intrigue the tulip has been involved in down through history. The "tulip-bulb mania" is perhaps the most unusual tale that we encounter in our brief exposé of this historical plant.

The origin of the tulip was in Turkey, and during the sixteenth century these beautiful plants were brought into Western Europe, where demand grew to a crescendo. The decade from 1630 to 1640 saw "tulip-bulb mania" prevalent in Holland with extravagant prices being paid for rare bulbs. Prices of as high as five thousand dollars were paid for the variety *Semper Augustus*, and tradesmen were willing to trade their homes and businesses for as little as one single tulip bulb.

Thereafter, bulb prices began to plummet and tulip haters began to spring up across Europe, Holland in particular.

In those ancient times, the tulip was still a novelty in Europe, and few had seen a tulip until a botanist, Ogier Ghiselin de Busbecq, visited Constantinople and saw a myriad of flowers. Among the familiar narcissus and hyacinths, de Busbecq saw one called "tulipam," so named by the Turks. The term was misconstrued by de Busbecq who thought it sounded like "thoulypen." Nevertheless, "tulipam" was condensed to tulip.

The tulip quickly diffused through Europe and was fondly received in Holland.

Tulips can produce their own rare types since cultivated varieties "break," and one can predict the new colors and patterns that are often extremely beautiful. Natural mutations ("sports") cause breaking of tulips as does a virus which spreads from bulb to bulb giving rise to its own cellular genetic message controlling the color inheritance of the bulb. As gardeners continued to cultivate the tulip, they realized that any tulip could break and yield a sizable fortune for the owner.

"Tulip-bulb mania" became an enormous phenomenon, and the gardens of Holland were radiant with tulips. People would speculate and gamble on as yet unbroken and unseen varieties, hoping to profit when the tulips bloomed. "Tulip-bulb mania" flourished.

Hollanders sold their antiques and family heirlooms, mortgaged their homes, farms and businesses, pawned anything of true value in order to raise money for their tulip dealings. Taverns, beer halls and even livery barns became tulip bulb exchanges, where dealers watched inflated bulb prices being pinioned onto boards.

By 1637, the tulip bulb market collapsed simply because traders and dealers grew weary of the game and released copious amounts of tulips on the market. Prices fell dramatically, and the reaction was analogous to the Crash of 1929 in the United States. The rich became poor, people committed suicide, and general chaos prevailed in Holland.

The Dutch government finally stepped in and prohibited further speculation in tulip bulbs; Dutch growers began raising tulips as a genuine business. Today, well over two billion specimens are grown per year.

Tulips are divided into several classes, all differing in time of bloom, flower shape, and size of plant. Over four thousand varieties of tulips have been named. Seven weeks of flower bloom can be expected by using the various classes—Early Single, Early Double, Mendel, Triumph, Darwin, Lily-Flowered, Parrot, Breeder, and Double Late.

Tulipa gesneriana is believed to be the parent of most of the modern tulips, such as the Darwins, Parrots and other hybrids. *T. gesneriana* has never been known in the original wild state.

Another species which has been cultivated for years is *T.*

Tulip
Tulpe
Tulipe
Tulipan
Tulipano
Tulipant

suaveolens, the tulip of the modern Duc van Thol. These tulips are characterized by their pointed "petals" and by their redolence.

Among the wild species which up to now have defied commercial breeding are included:

Tulipa clusiana—Quite often referred to as Madame Tulip and is indigenous to the northern portions of the Mediterranean shoreline. This variety will produce a slender flower on a relatively tall stem. Outer petals are rose-red and inner petals white. *T. clusiana* is well suited not only for Northern garden arrangements but can also be planted in more Southern regions of the United States.

T. greigi—A fine, early-flowering species, can be distinguished by its gray-green leaves with purple spots. Several forms exist; fiery red with a dark yellow-edged blotch at the base of each petal is the most common type.

T. eichleri—Similar to *T. greigi* in that it also has a large flower head with the petal bases having more of a medium-yellow shade.

T. acuminata—Has medium to light yellow flowers with noticeable red streaks. The petals are quite long and tenuous with the edges oftentimes curled backward.

T. biflora—Has a small white-tinted flower on stalks which branch. *Biflora* will bloom early in the season.

In modern times, tulips have been reserved for some of the most regal gardens in this country. Tourists have flocked to formal gardens like Dunbarton Oaks, in Georgetown, outside the District of Columbia; the Longwood Gardens, owned by the Dupont family and located near Philadelphia, Pennsylvania; and, surely the most important and famous gardens of all, the White House Gardens.

Tulips of the Rose and Jacqueline Kennedy Gardens

I. Darwin Hydrid
   A. *Empire State*—Tomato-red, large yellow base.
   B. *Gudoshnik*—Creamy peach-yellow, flushed-red, base bluish-black.
   C. *Lefeber's Favorite*—Glowing scarlet, base deep golden.
   D. *Spring Song*—Bright-red, flushed-salmon, base white.
   E. *Jewel of Spring*—Sulphur-yellow.
   F. *President Kennedy*—Golden-yellow, large bloom.

This new class of tulips, the largest developed as yet, is the result of crosses between Darwins and *Tulipa fosteriana*. No other tulips equal its size: large flowers, some stems reaching thirty inches in height. They are used as background in mixed border plantings or by themselves.

II. Cottage Tulips
   A. *Lincolnshire*—Geranium-red, base yellow.
   B. *Mother's Day* (Lemon Queen)—pale lemon-yellow, base deeper yellow.
   C. *Mrs. John T. Scheepers*—soft yellow, very large flower.

The Dutch call this group "Single Late Tulips"; other people call them "Cottage." This tulip has an interesting history, the older varieties having been found in cottage gardens in England. Many fine varieties have been added to the class by hybridization.

Cottage tulips have various shaped flowers and varying habits. They are splendid for gardening purposes and as cut flowers indoors.

III. Lily-Flowered Tulips

A. *Queen of Sheba*—Glowing brownish-red, edged orange.

B. *White Triumphator*—Pure white, beautifully reflexed petals.

C. *Mariette*—Deep satin rose, well-shaped flower.

Tulips of this class were once part of the Cottage class; they bloom simultaneously. In the last revision of the classification, however, they were put into a class of their own.

This class could be described as a tulip which does not look like a tulip. The shape is most distinct, resembling the lily. The long, beautiful, reflexed and pointed petals form flowers of great beauty.

IV. Parrot Tulips

A. *Black Parrot*—Glossy, purplish-black, strong stem.

B. *Texas Gold*—Deep golden-yellow, narrow red edge, large flowers.—Former First Lady Mrs Lyndon B. Johnson's favorite.

C. *White Parrot*—Pure white.

No wonder this class of tulips is becoming more popular. These large, artistically-shaped flowers with heavy, fringed petals have no peers in superb effect. Parrot Tulips are recommended for planting in clusters or in groups and for intermixing with other flowers.

V. Fosteriana Tulips

A. *Red Emperor* (Mme. Lefeber)—Blazing vermilion-red.

B. *Feu-Superbe*—Cardinal-red, edge yellow, base black.

These are the most brilliant of all tulips written and spoken about in this class. These tulips are among the group called "Botanical" or "species tulips." Among the *Fosterianas* is one of the most popular varieties, *Red Emperor*, which is unmatched for brilliance in size and coloration.

VI. Single Early Tulips

A. *Charles*—The color of this variety is scarlet with a yellow base.

B. *Ibis*—Deep-rose red, large-flowering.

C. *Pink Beauty*—Deep-rose with white flush outside, large and well-formed flower.

D. *Sunburst*—Orange-yellow striped and feathered red.

As this name implies, the varieties in this class produce single flowers and bloom early. Planting should be close to a building in shade or full sun.

In this class are flashing colors. The size of the bloom and length of the stem may not be as dominant in these varieties as in the May-flowering ones, but their presence in the garden as forerunners makes them desirable. Ideal for bedding or clumps, they can be used effectively in front of a border which is planted with taller kinds.

VII. Mendel Tulips

A. *Bing Crosby*—Rich, glowing scarlet.

B. *Sulphur Triumph*—Primrose yellow.

Mendel Tulips are important in that they bridge the gap between Single Early and Darwin types. Mendels bloom after the Single Early and about two weeks before the Darwins. They have colors and substance of Darwins and also the earliness, because the Mendels are crossed with the Duc van Tol. Mendels can be used for clumps or bedding.

VIII. Triumph Tulips

A. *Blizzard*—Creamy white, large flowers.

B. *First Lady*—Red-violet, flush purple.

C. *Hibernia*—Pure white.

Flowering after the Mendel Tulips, this class continues the blooming succession. With Mendel types, this genre represents the midseason blooming time. The Triumph class originated from crosses between the late-flowering Darwins and late-flowering varieties in the Single Early class. Triumphs have large flowers and strong stems and range in medium height. They are for bedding purposes.

Additional White House Bulbs

The ancestor of the present-day hyacinth is *Hyacinthus orientalis*, which was quite common during ancient times as a wild flower in parts of the eastern Mediterranean and throughout Asia Minor and Syria into Persia.

The true hyacinth has a blue flower; however, centuries ago the hyacinth developed other color variations, such as white and purple. In the 1560s the hyacinth was introduced into Holland where it has been cultivated ever since.

One reason why the Netherlands has become world-famous for hyacinth production is because of the strict control of the water table in the ground. Holland's bulb district is only eight inches above mean sea level and is protected by dykes.

Another reason why the hyacinth is par excellence is because the sandy soil is heavily manured and extreme care is taken during growing.

Hyacinths are exported to the United States in six grades:

Exhibition—These are the largest bulbs that are grown. They bear the largest flower and bulbs are nineteen centimeters in circumference.

First Size—These bulbs range between eighteen and nineteen centimeters.

Second Size—These bulbs range between seventeen and eighteen centimeters.

Third Size—Ranging between fifteen to sixteen centimeters, these bulbs are excellent for pot forcing or used in the garden.

Bedding Size—Muscari grape hyacinth, the size used for borders in the West Garden of the White House and other locations on the Executive Grounds, popular for outdoor planting. Can withstand wind and rain. Bulb size similar to Third Size.

Miniature—The smallest of the hyacinth which produce fine flowers.

Allium

Allium is the ornamental cousin of the onion. Chives are also in the onion family. Two kinds of Allium are popular for the garden and easy to handle. The low-growing Allium reach a height of less than one foot, and the giants reach five feet. Both types have large, round flower heads called "umbels" and are ornamental. Some Allium are fragrant and some emit an onion scent.

Allium planted in the Jacqueline Kennedy Garden, in particular, add accent to the spring bulb display and last some time after the tulips.

Of the selection of Allium to pick from, *Allium albopilosum* was quite popular in the Presidential Gardens.

Chives

Chives are bulbs in the onion family and have served two purposes in the East Garden. Since chives are perennial, they were already in the East Garden display when I arrived at the White House. I occasionally replaced chives as time and attrition dictated. In each of four boxwood-enclosed garden plots, we used chives to give variety of color, shape and utility to the garden. Chives are hearty growers and have reached heights of seven to eight inches before either I or Chef Haller would snip the plants back. Chef Haller would use the chives in the White House kitchen as a spice.

Daffodils

Daffodils will grow in most any part of the United States and in most any type of soil including the White House soil medium. Some people may ask what the difference between daffodil, jonquil, and narcissus is, and the answer is that daffodil is the common name for *narcissus*, the species. Jonquil is the name of one group of narcissus and its hybrids.

Additional emphasis is given to the term "daffodil," since there is an American Society for the Daffodil, and one in England, both of which have daffodil shows and yearbooks.

The daffodils that are hidden behind the osmanthus hedges of the East and West Gardens are the Trumpet type and are called King Alfred. John Kendall created this type in 1899, and it is considered the greatest advance in the cultivation of daffodils. It is the most widely grown type.

White House Plant Pests and Diseases

I. *Aphids* are white to yellowish-green insects about one-eighth of an inch long that settle on the undersides of leaves and on plant stems, buds, and stored bulbs. Aphids suck the plant juices, causing leaves to curl and flowers to deform. These insects also spread virus disease and secrete honeydew, a sticky substance that attracts ants. Some of the more susceptible plants include dahlias, gladiola, irises, roses, and lilies. Methods of control include knocking off heavy infestations with a stream of water from a garden hose. Infested plants should be severely sprayed with malathion, nicotine sulfate, or a heavy solution of soapy water. Undersides of leaves should not be neglected.

II. *Gladiolus thrips* are insects one-sixteenth of an inch in length with bristly wings. Thrips are yellow when young and brown as adults. While on leaves they cause streaks that appear silvery at first, turning brown later. Thrips also streak and wither edges of flowers and cause stored bulbs or corms to become sticky and brown. In winter, thrips breed at temperatures of 60 degrees F. or greater. The main goal of thrips is gladiola, but may also attack freesias, irises, and lilies. Control of thrips includes spraying with sevin, or diazinon; begin the spraying when plants are six inches high. Dry and spray dug bulbs before storing for winter.

III. *Japanese beetles*, bronze colored and up to one-half inch long, will chew on foliage, flowers and stems thereof, causing large round or irregularly shaped holes in leaves. Japanese beetles are more active on warm, sunny days, when they can be seen in clusters munching through one plant and moving onto others. Cannas, dahlias, roses, Rose of Sharon, and wild rose are among their favorite plants. The beetle can be controlled by individual or group picking them from plants and depositing them in a can of kerosene. More heavy infestations can be sprayed with carbaryl or methoxychlor.

IV. *Slugs* are fat and legless, grow from one-half inch to five inches long and may be gray with dark spots, yellow-white, brown, or black. They are mostly nocturnal feeders but are seen during early morning. Slugs eat ragged holes on foliage near the ground and leave a telltale silver slime trail on branches and sidewalks. Dahlias and lilies again are prime targets. Slugs can be controlled by dusting or spraying the soil with metalclyde or with metaldehyde pellets. Snailrol was commonly used at the White House; the slugs would roll around and through the fine particles of pesticide during nightly forays into the gardens.

V. *Stem and bulb nematodes* are microscopic, transparent wormlike animals causing plant leaves to twist, split, and blotch in yellow and brown colors. In severe cases, foliage may fail to appear. Nematodes enter new shoots and go upward, later to retreat to the bulb as plant dormancy approaches. The bulbs, therefore, turn brown. Bulbs that are affected include daffodils, hyacinths, irises, snowdrops, and tulips. Infected plants should be destroyed to prevent spreading of the infestation. New plants must be set out in a different location, or the affected soil should be fumigated with a recommended nematocide.

VI. *Bulb mites*, which are quite small, about one twenty-fifth of an inch long, have eight legs and resemble little spiders. Bulb mites colonize to feed on rotting bulbs, but may also tunnel into healthy bulbs. In this manner, the malady known as bulb rot is transmitted. Tulips, lilies, hyacinths, and crocuses are suitable for bulb mites. In controling bulb mites, throw away rotted bulbs while dusting good bulbs with dicifol. Store bulbs in tightly closed containers.

VII. *Mosaic* is a virus affecting hyacinths, irises, cannas, and daffodils among others. It is transmitted primarily by insects such as the aphid and may cause leaves to become mottled; it stunts and distorts leaves and flowers. The "tulip-breaking" mosiac has given rise to the multicolored blossoms of the Rembrandt tulips. Infected plants should be destroyed. Mosaic cannot be chemically cured, but the mosaic transmitter—aphids—can be sprayed with nicotine sulfate or soapy water. Tulips and lilies must never be planted together since the virus that adversely attacks lilies also attacks tulips.

VIII. *Botrytis blight* will produce yellow, orange, brown or red-brown spots on leaves, flowers and bulbs, especially on rainy, humid days. Spots tend to grow together, causing affected plant parts to fall and become slimy with moldy cover. Tulips, dahlias, hyacinths, and tuberous begonias are usually affected. Those bulbs having round brown lesions should not be planted, but destroyed. Remove diseased plants and change planting location for new plant specimens. Plant in well-drained soil and in full sun. Space bulbs to increase air circulation and mitigate humid conditions. Sprays to use are phaltan, maneb, and zineb.

IX. *Bulb rots* produced by bacteria and fungi will bring about yellow, stunted shoots with brown, mushy, or dry, spongy bulbs. Rots are particularly destructive in poorly drained soils in warm, moist weather. Bulb rot is a scourge on amaryllises, calla lilies, dahlias, tuberoses, and tulips. Infected plants and bulbs must be destroyed. Set out new bulbs in well-drained soils. Choose a sunny spot for planting bulbs, and provide generous spacing to increase circulation of air. Store bulbs in dry, well-ventilated places, and dust accordingly.

X. *White flies* are small, snowy-white bugs, which usually infest a plant by the hundreds. These creatures light on the underside of plant leaves and damage the leaf structure to a point whereby the plant begins to droop. White-fly infestation comes in cycles, and so also should the cycle of spray treatment. Malathion has proven effective against this pest, but may need to be rotated with another insecticide.

Annual Flowers at the White House

Geraniums *(Pelargenium)* were frequently used in the East Garden, West Garden, North and South Fountains and by the South Portico during the summer season. The geranium produces continuous blooms throughout the summer months and can be held until the first week of October, although the geraniums were removed from the White House gardens prior to this date.

Geraniums should be planted in full sun after May 15, when danger of frost has passed. A loam-soil medium is best for geraniums that provides adequate drainage. Watering is done early in the day so that the plant can utilize the moisture before the latter portion of the day; never water plants at night. Watering is done from under the plant in hot weather so as not to ruffle the plant or the water possibly burning the plant as it rests on leaves on the sunny side. Because of the short growing span of the geranium in the White House gardening scheme, prefertilizing the beds was sufficient to sustain the plants until the time of removal in favor of a fall display. Otherwise, a liquid fertilizer applied once a week would be needed. Propagation occurs by taking cuttings, which can be rooted in perlite or sand mixed with Michigan peat moss. Geraniums can also be propagated from the parent plant and left in a cool, dry place until the cuttings begin to callous. Thereafter, place the cutting in a propagation medium. Although geraniums produce prolific blooms, blooms turn brown and were removed from the plants once or twice a week from White House flower beds. This necessitated walking carefully through the beds and tossing faded blooms into a basket on the lawn. Fertilizer creates verdant growth of leaves, and to increase air circulation, leaves are pinched off while faded flowers are removed. Favorite geranium varieties used at the White House include Improved Richard, Rose, and, in the rooftop hothouse, Ivy Geranium is favored.

Wax Begonia *(Begonia semperflorens)* is one of the most popular and easily grown of the common garden plants. Begonias are almost untouched by insects and diseases and require moderate watering and sun for healthy growth. Fertilize moderately. The begonias can be pinched back when they become overgrown, and the plant comes forth again with lush growth.

The Wax Begonia does well in the East Garden of the White House, and, because of its rather short stay in this garden arrangement, fertilizing was not needed. The begonias have suffered occasionally, however, during extended hot weather; I then simply ordered new plants from the Kenilworth greenhouse as replacements. Pink and red varieties of begonias were commonly used in the East Garden, while the white variety was used in the flower shop of the White House. Begonias can be perpetuated by either seeds or cuttings, but my experience has shown that raising seeds under properly controlled hothouse conditions far surpass cuttings.

Marigolds (*Tagetes erecta, etc.*) consist of a very large and complex group of annual flowers. Because of their ease of propagation and popularity, plant breeders have developed new varieties. Those marigolds of the South Fountain displayed during 1973 had a disagreeable odor, but many new types, thanks to the breeders, have been developed without this smell. The marigolds' habit of growth varies from the tall, bushy American groups to the dwarf, multibranched Mexican plants. The French marigolds are stocky with bushy growth. General care of the marigold should include watering during the summer heat spells if they are to continue their prolific blooming. Cultivating with a light garden fork should be done carefully and made shallow, since marigold roots are not deeply entrenched, and ripping up a plant or two is not hard to do. If one can endure the smell during a hot, still day, then blooms that are spent can be picked from the summer display. I can remember how sick I felt when having to commence with this duty at the South Fountain. To curtail vomiting, I had to desist from mingling in the marigold bed altogether.

Marigolds are good for bedding purposes, as cut flowers, and as background support. The plant blooms from June until frost, and the varieties grow from six inches to three feet. The soil medium can be loam, and exposure, to insure good blooming should be full sun.

Marigolds, like begonias, have little trouble with insects or disease, but Japanese beetles may occasionally feed on blooms. Such injured specimens can be pulled and discarded.

Nicotians (*Nicotana alata grandiflora*—flowering tobacco) were used sparingly in the West Garden, primarily as an accent plant in the evening. It is during the night when the starry flowers are open and the fragrance floats across the garden that the plant is most appreciated. The flowering tobacco will self-seed, but most seeds thus generated are of little value to gardeners.

The older varieties of the flowering tobacco were slender and tall but the newer types are shorter. The color spectrum ranges from white through red, maroon, rose, and yellow. We used a yellowish-green variety in the West Garden.

The pansies (*Viola tricolor hortensis*—Pansy heartsease) were most commonly used as a fall-to-spring plant in the East Garden together with spring bulbs. The White House method of planting and care included planting at the same time as tulips and mulching with pine branches and bowls to guard against frost heave during the winter. Pansies were one of the attractive perennial plants that we treated as annuals in the East Garden. Pansies are cool-weather plants so that plantings in the autumn will produce flowers in the plants by March of the following year. Flower production in early spring makes a marvelously effective border or bed. The pansy is inexpensive and available in many colors. They benefit greatly by pinching back as scraggly shoots develop. The more pinching, the better is the succeeding bloom. Faded blooms should be removed promptly.

Other gardening particulars for the pansies are height, which is anywhere from eight to twelve inches. The soil should be rich and moist as the East Garden is. Blooming from March to June, pansies make good cut flowers, and the exposure is sun and light shade.

One of the popular varieties used in the East Garden was Pay Dirt, although many types exist in the pansy group.

Another annual favorite of the White House Horticulture Staff is the petunia. (*Petunia hybrida*). Petunias are currently one of the most popular annual flowers in the United States. Petunias are relatively resistant to insects and can withstand drought conditions. They are profuse bloomers and can be used in beds, edging, windowboxes and standard tubs. The latter use was frequent at the White House, and they performed well throughout the growing season.

When one is planting petunias in outdoor gardens, the soil should be warmed and there should be no danger of frost—after May 15 in the Eastern United States. Soil medium should be enriched with compost, leaf mold or peat moss, the latter additive being used for White House plantings during my tour of duty. As petunias continue to grow, they become long and stringy, and it is advisable to pinch the plants back to stimulate new and vibrant growth and to prolong blooming. A prefertilized soil is often suitable for sustaining nourishment of petunias for garden plots, but I strongly recommend one or two additional feedings for tubbed or potted petunias.

Petunias, because of their present popularity, have many varieties to choose from, and more are being added to the list every year. Some of the varieties that I have raised at Kenilworth and which were planted in wooden tubs and garden beds for the White House include Comanche, Capri, Mercury, and Apollo.

Most of us on the White House staff did not pay much thought to the purple coneflower or the black-eyed susan (*Rudbeckia*), the latter having been used in the East Garden as well as the West Garden during the later 1960s. The *Rudbeckia* represent a hardy plant which can survive our climate, withstand heat and drought, and maintain bloom throughout the summer. The hybridizers have taken advantage of the tough native *Rudbeckia* and have developed plants with colorful bloom and sustaining flowering. The *Rudbeckia* are subject to few insect and disease blights.

Rudbeckia planted in the gardens were done so in full sun, although these beauties can tolerate one-half sun or a little less.

They form bushy plants, which were what I needed in the East Garden display, and, depending upon type, grew from one to three feet tall. To be cognizant of new varieties, I used the Gloriosa Daisy and Gloriosa Double Daisy—both over two feet high—in the East Garden.

The *Rudbeckia* are especially nice for cut flowers, and Rusty, who worked in the White House flower shop, used them quite often for the main house. Wind and rain will push the taller types of *Rudbeckia* over, and drooped flowers require staking with cord and bamboo to prevent plant stems from breaking.

Zinnias (*Zinnia elegans, Z. angustifolia*, etc.) are another of the popular annuals, simply because they require minimal care; indeed, they are a lazy man's flower. Zinnias bear color and bloom for the entire summer period, and they have strong stems, heavy foliage and flowers, and present themselves in an effective scheme.

The primary place for setting out and planting zinnias was in the East Garden along the glass-enclosed colonnade, where thousands of tourists shuffled by weekly, watching my progress in the garden and eyeing my flowers. Although zinnias maintain healthy foliage during the opening months of summer, August brings cool nights and mildew to the zinnia leaves. Thus, the plant begins to turn brown and, with autumn breezes in the air, are pulled from the garden one after the other.

Zinnias comprise so many types, strains, and varieties that the gardener may become dumbfounded by the selection. The giant dahlia-flowered zinnia is used by some gardeners over the older flat flowers. The cactus-flowered zinnia with its quilled petals is also favored. The small-flowered zinnias—Lilliput, Pompom, Baby, and Cupid—are as pretty and useful as their larger-flowered relatives. Other forms include Crested and Double-Crested zinnias and are quite interesting in form and color.

A rich soil is a prerequisite as is an adequate amount of moisture for strong, rapid growth. A prefertilized soil will serve to enhance initial zinnia growth, but supplemental feeding during the summer will help the zinnias to look their best. After all, the East Garden is the First Lady's garden, and I would hate to have to be called on the technicality of ugly zinnias.

Zinnias can suffer from several insect pests and diseases. Japanese beetles attack zinnias, and mildew is the most common disease. Malathion can control the beetles, and a sulphur dust the mildew.

Although dusty miller (*Primula auricula*) does bear a small, round, yellow flower if left all to itself, the flower is pinched off; the plant is grown for its foliage and used in borders. Set out after May 15 when the soil is warm. This interesting plant gives beautiful color contrast when used with red geraniums. The color of the dusty miller is, as the name implies—a dusty hue, like powder covering the plant, and from certain angles, it has a rather unique ghostly charm all its own.

Dusty miller plants are best raised from cuttings, which are taken from the parent plant and placed in a perlite/peat-moss propagation bed. The dusty miller is one of the easiest to propagate and a hearty grower in the garden.

The dusty miller plant is best used in a two or three staggered set of lines and lends itself quite well to a typical hedge-beveling operation. The above operation was commonly done under the flowering Malus of the West Garden summer display (diamond-shape cut) and under the topiary holly of the East Garden summer display (square cut).

Dusty miller requires an enriched soil with adequate watering and must not be left to flower, because "going to seed" saps the strength of the plant. The flower head is too small to be of any aesthetic value. The plant can be left as a bedding display from May until frost or, as we did in the Presidential Gardens in 1972, until frost, when we changed the bed from dusty miller to the silver-gray herb Saint Helenia.

Another favorite plant of many gardens and of the White House gardens is the perfume redolence of Heliotrope (*Valeriana officinalis*) with its amethyst-colored flowers and lush, green foliage. Heliotrope are suggested for border plantings, tub culture or as single specimen plants punctuating the summer garden. Because of its agreeable fragrance, the heliotrope was in demand as a cut flower in the White House flower shop. Outdoor planting in the summer garden should be done in a loam soil, and the plants can be watered early in the morning, and, if leaves droop during the day, water again before dark. Heliotrope require full sun for optimum development, but can tolerate up to half-sun. These plants grow best in soil pH* of about 6.5 as do most plants so far described. The heliotrope plant is a rather quick grower with topheavy blooms that may topple the plant during rain and strong wind. In such case, the plant can be staked to bring it upright, or, if so desired, flower clusters can be snipped to allow the underfoliage to develop. It had been my experience at the White House to cut topheavy heliotrope flower clusters and allow embryonic buds on the stem nodes beneath the flower head to come to full blooming cycle. This treatment provided good symmetry to the plant and stimulated roots and flowering.

Heliotrope can suffer miserably from white-fly infestations, and a program of malathion treatment can help abate this scourge. A test to determine white-fly presence on heliotrope is to slap the plants and see how many white flies take off into the air around the specimen. If the plants are particularly in need of a spraying, then wait an hour so that the white flies once again alight on the plants and commence treatment. Some seasons are unusually bad for white flies, as is also true for aphids or beetles. The summer of 1967 was dubbed the "year of the white fly" by me because of the huge volume of this pest at Kenilworth. At the time, I thought I would never see the end of the albino creature. Spraying never seemed to work, even with spray rotation.

*pH. A term used to express the acidity or alkalinity of soils and solutions. It is the reciprocal of the logarithm of the hydrogen ion concentration. (H^+) expressed in gram-ions per liter, therefore:

$$pH = \log 1 \log H^+$$

A neutral soil has a pH of 7. Some common pH soils range from 3 to 7. Most plants can tolerate a pH range of 5.9 to 7.0

Perennial Flowers at the White House

Besides using candytuft (*Iberis saxatilis, I. sempervirem*) in the East Garden as square borders between topiary holly, I found this plant had little use elsewhere on the Executive Grounds.

The perennial candytuft, one of the best of edging plants, can be described as a small shrub. It is very much in demand for rock gardens, is of easy culture, and can be raised from seeds, scions, or divisions. These plants can form a dense mat of foliage and, once established, should not be disturbed. The candytuft is fairly low growing.

Candytuft is suitable for windowboxes as far as the East Garden was concerned, since four redwood boxes of the plant were placed over a manhole on the south side of the garden to complete the flower scheme. In such culture, however, one must keep an eye on the planters so as not to permit the candytuft to dry out and begin to droop from lack of moisture.

The candytuft blooms from April to June and attains a height of from six to twelve inches. Cuttings are taken in September as is division. The plant will seed in August. Candytuft requires a moist, moderately rich soil as well as full sun.

Pruning is optional, and on occasion I did prune my candytuft in the East Garden to separate it from the herbs and annual plants.

One of the most showy of all autumn plants is the chrysanthemum (*Chrysanthemum idicum, C. morifolium, C. sebircum chrysanthemum*) or mum for short. It has been the reenforcer of all autumn perennial flower borders and in recent times have given us many early-blooming varieties in myriad types and colors which survive the first frost. The flowering season is tricotimized into early, midseason, and late: respectively, September 5 to October 5; October 5 to October 25; and October 25 to November 5. The color spectrum varies from pure white through the yellows, pinks, bronzes, purples, and red. All have been used in the fall display in all of the White House Gardens.

Mum buffs speak of the different varieties by bloom structure. Two visible characteristics are identified by difference in the outer petals and disk petal centers. A few of the major classes of the modern chrysanthemum are singles, anemones, pompoms, incurves, decorative reflexes, spoons, quills, and spiders.

Division is one method of propagating mums in the spring. An even better propagational method is rooting cuttings, which are taken from the top three inches of the newer shoots when the plants are at least eight inches high. This was done at the Kenilworth greenhouse in preparation for the dozens used on the White House grounds. The propagation medium is a moist mixture of sand and Michigan peat, and cuttings can be expected to make roots in ten days to two weeks.

At the end of the blooming period, the plants should be cut off four to six inches from the ground and the tops discarded. Using a light, loose medium such as peat moss, mulch the mums after freezing of the ground to prevent recurring cycles of freeze/thaw.

Although the above exposé on mums is true and wanting, the White House used the chrysanthemums as if they were annuals. They served for a short time only, between the summer display and the beginning of the tulip season. Thereafter, the mums were torn out of the gardens and were distributed to interested parties for their use the following season. The West Garden did, however, have mums which were used as year-round specimens.

Chrysanthemums require a rich, moist garden loam, full exposure to sun, and should be planted twelve to eighteen inches apart. Mums may grow from one to five feet and were used for cut flowers at the White House as well as individual plantings.

Chrysanthemums used at the White House were of several varieties:

1. *Jesamine Williams*, which is a white quill, maintains medium height and blooms in early October. It was used in the West Garden.

2. *Red Mischief* is a red mum, a single type, is slow-growing and is in bloom by mid-October. It was used in the East and West Gardens.
3. *Yellow Supreme* is naturally yellow and is a decorative, low-growing early-October mum. The South Fountain was planted with this variety.
4. *Pink Chief* is an incurve, medium-height chrysanthemum, blooming October 15 to around the October 23. It has been incorporated into the East Garden.
5. *Red Headliner*, decorative among mums and grows to a medium height, blooms by the latter days of September. It was used in the South Portico Garden.
6. *Bronze Song* is of an autumn bronze color and is a decorative type and reaches medium height and blooms in mid-September. It has been used in the North Fountain flower bed and inter-mixed in the West Garden.
7. *Indian Summer* is red, a reflexing type, tall and in bloom by the third week of October. It may be used in the East Garden and other locations.
8. *Pumpkin* is bronze in color and a reflexing type that is short. Pumpkins bloom just after mid-September.
9. *Ostosa* is a fine white mum and of decorative type. It is a low grower and blooms around the first week of October. It was used in the West Garden.
10. *Dolli-ette* is a beautiful bronze mum of the quill type. It is of medium stature and blooms in early October. It is one of my favorites in the Presidential gardens.

One of the rarer perennials to be found in the West Garden of the White House, the columbine (*Aquilegia* species), is nevertheless a dainty, year-round plant for the early spring garden. The newer types of columbine, with their long, spurred flowers, are beautiful and available in a color spectrum suitable both for cutting and border display. Columbine cannot tolerate hot sun and, therefore, will need protection.

Canadensis, which attains a height of eighteen inches, has red-orange flowers, while caerulea, which was used in the West

Garden, grows to two feet and is the famous Rocky Mountain blue columbine.

Columbine bloom from April until June and reach from one foot to three feet in height. They need to be planted in early spring or in September and can take half-sun.

Plant columbine seeds in June when the soil is warm and moist, and divide columbine in September before replanting.

Columbine is often bothered by leaf miners, which do little damage other than disfiguring foliage.

In 1969, we added several daylily (*Hemerocalli*) plants to both Presidential Gardens, and they grew very well in their new environment. The lilies are quite dependable among the perennials, being vibrant, tolerant, free-blooming, and immune to many insects and diseases. The daylily requires little care and practically no fertilizer. Water the daylily while it flowers.

Some daylily varieties are dwarf, while others are tall—approaching four feet. The latter specimen was used in the White House gardens. Daylilies function best in full sun, but will waste away in shaded plots.

Daylily coloration has been the usual yellow, but pinks and reds are being developed. I preferred the traditional yellow for my summer garden complement on the grounds I don't believe pink would be attractive in many schemes.

There are, in addition to the varieties above, numerous new ones, which include double-flowered types, miniatures for smaller garden plots, evening-flowering varieties, and late-flowering varieties. Too, there are reblooming, continuous blooming and variations in flower, color and form.

Daylilies bloom from May until September, reaching anywhere from one and a half to four feet, which gives them application in both border and bedding. We used them as single specimens. The plant can be propagated from seed and division, seed in the greenhouse in November, and division after flowering. Plant the daylily April through September in garden loam replete with humus.

Lily-of-the-valley (*Convallaria majalis*) is one of the most fragrant of the perennials and has been used in the Children's

Garden simply because of shade conditions and difficulty in plant selection for that area. This plant is a spreader, and thought must be exercised when planting it. Lily-of-the-valley plants like a soil enriched with leaf mold, compost, or peat moss and, if available, cow manure. Plant so that roots are at least two inches below ground level and six to ten inches apart. Do this preferably in the fall or in March.

Lily-of-the-valley is grown for its cut flowers, for ground cover, and for fragrance. Blooms commence from May to June, and the plant does best in shady areas. The plant is seldom bothered by insects or diseases.

The peony (*Peaonia officinalis*), used sparingly in the West Garden, is one of the older spring-flowering perennials. It is good for borders or for single specimens. Peony plants offer wonderful cut flowers, although their flower heads are exceedingly large for many displays.

Peonies will last for many years; therefore, their beds should be looked after prior to planting and post-planting. Set the peony roots with the eyes not over two inches under ground, or they may not bloom.

Supplemental fertilizing is essential for best results and bonemeal is one of the better feeding media to use.

Peonies may suffer from budrot, and this malady can be controlled by brushing off the agents causing the rot—ants and aphids. Also remove old foliage by autumn.

Tree peonies, one of which we had on the South Grounds near the southeast fence, require light shade, unlike their cousins, which must have full sun. The tree peony can be planted up to six inches deep; they will retain their above-ground stems, which are easily broken if not careful.

Pachysandra (*Pachysandra terminalis*) or Japanese spurge, was one of the most used ground covers at the White House. Especially prevalent intermingled with the hollies near the South Fence, it is a shade-lover and is used to cover areas where grass refuses to grow. The prepared soil for pachysandra should be replete with aged manure or compost, and a top dressing of same should be done after planting to keep the roots cool. On small cover plots, the flowers of pachysandra can be pinched off as well as the plant tips themselves. Pinching the tips encourages root makeup and extension.

Pachysandra usually is not attacked by disease, but treading on the plants tends to enhance a leaf disorder, which can be remedied by removing the infected area.

Pachysandra desire well-drained loam as a growing medium and can be planted one foot apart, although I have placed them four inches apart in filling voids in beds along the fence on the South Grounds. These cover plants can be propagated by division in April and September and by cuttings in July and August.

Quite abundantly planted in the Children's Garden between the flagstones in 1968, *Vinca minor* (Periwinkle, Running Myrtle) is one of the most useful of the ground covers. *Vinca minor* is an evergreen and grows best on shady slopes. It may grow under some sun, but will grow more slowly in the sun.

Vinca minor can be set out in early spring or, as we did in the Children's Garden, in early fall. Set out plants about ten inches apart, and, in the course of one or two years, by the mechanics of plant layering, *Vinca minor* will form a long-lasting mat.

Vinca minor will grow up to eight inches and will bloom in April and May. Propagation of *Vinca* is by division and cuttings in April or September. Ordinary garden loam will suffice for growing medium. Varieties of *Vinca* include Bowles—a deep blue; Minor alba—a white; and Miss Jekyll's—a white.

The White House Lawn

The White House lawn, not including the house and certain patches of trees and shrubs, comprises just under eighteen acres, and maintaining this expanse of grass entailed much work. Kentucky Bluegrass (*Poa pratensis*) was the mainstay for all of the grounds, including the new 1972 charge, Lafayette Park, across Pennsylvania Avenue. However, K31 fescue was

Thyme

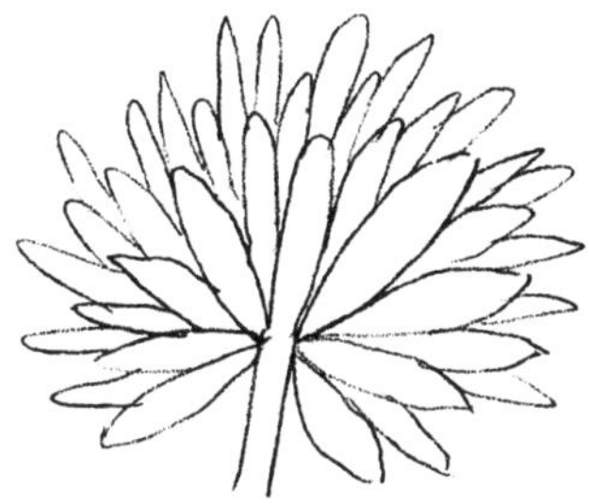

Chrysanthemum

used in the East and West Gardens during the winter of 1968, 1969, and 1970 onward. Summer cover for the above exceptions was a Zoysia grass (zoysia japonica), which was green at the Kenilworth greenhouse by late May and transferred and sodded at the White House in June. An excellent summer grass, Zoysia does become dormant and brown by the first or second frost, and if one can tolerate the ugly expanse, then so be it. Not so at the White House, as the Zoysia was removed by September 23 and the winter cover reset. The change of grass can be analogous to changing summer and winter rugs: the grass was cut, removed and new grass sodded.

Particulars for *Poa pratensis* include the following: Soils should be moderately acid, pH of 6.0 to 6.9, well drained and fertile. It needs nitrogen and phosphate; 5-10-5 can be used or the old standby, Weed and Feed. Cut grass so that over two inches are left, and never cut lower than one and a half inches. Seed at the rate of one pound per 1000 square feet, and seed in the latter part of August. Use a grass catcher to collect clippings during cutting. Fertilize at the time of seeding, and, after the grass is established, apply fertilizer once in the spring and fall.

Grass maintains good color and, if injured, heals quickly. It may become dormant if summer weather is hot and needs to be thatched and aerated when sodbound. Thatch in August at time of seeding.

The Zoysia japonica grass is tolerant to a wide range of soil types, but does not adapt well to sandy soils. When cutting, lower to one inch and no more. Zoysia is good for lawns and has the advantage of being dense, tough, and a long grower; it is wear resistant and can grow in light shade. Some of the disadvantages include dormancy in winter, slow establishment, and need to be planted vegetatively. Plant by plugs or sprigs, which should be two to four inches, and plant in mid-May to mid-September.

Fertilize during sodding and one feeding per year after planting. Lime to maintain pH of 6.1 to 6.6.

Although our Zoysia at the White House was segregated, Kentucky Bluegrass can be mixed with Zoysia, which will persist for a year after the Zoysia is established. Thereafter, the Bluegrass will choke out as will most weeds.

Flowering Shrubs at the White House

The azalea (*Rhododendron species*) is one of the most widely planted shrubs in the Washington area, and the White House can boast of having several fine beds. Our climate favors this most beautiful plant, and for the most part azaleas bloom in the spring with a wide spectrum of growth habits.

There are several native species, but there are hundreds of varieties and many colors to choose. Most of what are planted are Kurume group, and a large quantity of these are of the Henodegiri variety, a bright red, but sometimes with a magenta overtone.

Hnodigiri is one of many Kurume varieties, which produce such colors as the white variety, Snow, Coral, Lavender Queen, and Hinomayo. They are moderate in size and produce a mass of bloom each spring.

The next group, the Glenn Dale azaleas, were produced by the former director of the National Arboretum, B.Y. Morrison. Over four hundred varieties of the Glenn Dale group have been introduced into the flower business. These have stronger foliage than the Kurumes and larger flowers and at maturity they may reach ten feet in height.

Glenn Dale azaleas range in color from white to deep purple, and flowering is later and longer lasting than the Kurumes.

Another of the evergreens is the *Kaempferi*, or torch azalea. It grows to five feet and, because of its orange-red color, is of interest. Some varieties of this are Canmen, Cleopatra, Louise, and Othello.

Many of the earliest azaleas to bloom are the deciduous types. Varieties are Royal Azalea of Japan and the indigenous Pinkshell (*vaseye*). Other natives include Pinxter (*R. nudiflorum*), swamp azalea (*R. viscosum*), and the flame azalea (*R. calendulaceum*). The European Mollis hybrids, which we cultivated at the White House, included showy yellow to orange colors.

Azaleas grow best in an acid soil, one having a pH below 6.0. To decrease the pH in the azalea bed, finely ground leaves and moss can be worked into the bed. When planting azaleas, add much peat moss, especially near buildings, since concrete will increase pH from leaching into the soil. Acidity can also be increased by applying sulphur and permitting this element to dissolve and percolate into the azalea bed.

If azaleas continue to look yellow-green in the foliage, then there may be a deficiency of iron, which can be supplied in several forms. Iron chelate was commonly used at the White House, but micronized iron, or sulphate of iron can also be used. Apply iron chelate directly to the soil, not on foliage. Iron sulphate can be sprayed on foliage to help the situation.

Azaleas thrive in either sun or shady conditions; in the shade, flowering will last longer. If planted in the shade, they must be placed away from shallow-rooted trees, as the azaleas cannot compete with these neighbors. Mulch azaleas in the fall with pine bark so as to protect shallow roots from winter freeze. Azaleas can be mulched in the spring with a leaf mulch to provide additional nourishment and to maintain pH levels. Azaleas can be pruned after they bloom, and a spray for red spiders and lacebugs can be applied season long. If azaleas suffer from petal blight, a disfiguring disease, remove winter mulch by spring and apply new mulch. The petal blight winters over in mulch. Spray several times after replacing mulch and during the booming season when the weather is clear. Use additional spray during wet weather.

Occasional injury may occur to azaleas in the winter season, especially from cold, dry winds after a temperature drop. The bark on azaleas may split from low temperatures, causing the plant to die. The best protection offered for azaleas is to select strong, healthy plants, to plant in well-drained, aerated soil with a low pH, and to keep the plants moist during drought. Mulch plants before winter, and spray on schedule to lessen insect damage and disease.

One of the more beautiful flowering plants at the White House is the camellia (*C. japonica, sasanqua*, and varieties), all of which grow on the south side of the mansion. Many people believe that the camellia cannot be grown as far north as Washington, D.C., but, with proper planting and care, the camellia will thrive here.

The camellia is especially good for shrub borders, foundations, cut flowers, and as a specimen plant. It is the latter use that this plant saw on the White House grounds. The camellia can be planted from March to October, although I prefer the early spring so that the roots of the plant can become established during the summer months. Our camellias at the White House grew approximately five to seven feet high and we had both spring and fall bloomers. The Camellia japonica flowers in the spring while *Camellia sasanqua* will flower in September until the end of the year. Soil mixture for the camellia should be replete with leaf mold, compost or peat moss, and, as with the azaleas, the pH of the soil should be around 6.0. The exposure of the camellias on the South Grounds as well as the camellia by the South Portico should be light shade.

Mulch camellias as you would azaleas, using pine bark or pine needles. This provides the necessary protection from temperatures in extreme and helps maintain moisture content in the soil near the plant.

Prune camellias as they become larger and broader. Suckers appear each year and need to be cut in the late fall along with those of holly and orange.

While little spraying of camellias for insects was done during my time on the White House staff, this plant requires little in this area of the country. The camellia is also relatively free of such diseases as leaf blights.

The Pyracantha (*Pyracantha coccinea lalandei,*) or firethorn, was introduced into the landscape scheme of the Tourist Entrance for a second time in 1973, just after I returned to the White House. The original plants had to be cut out because of age and ugly appearance.

The Pyracantha is one of the more attractive evergreen plants and is useful in the shrub border, against walls or as used in the East Annex as hedges. They bear splendid evergreen foliage, colorful fruit, and are in demand for interior decoration.

Never overprotect the firethorn to a point where the wood fails to ripen early in autumn and will be subjected to winter freezes.

The firethorn is best planted in spring. It attains a height of six to twenty feet and blooms inconspicuously in late spring. The fruits of the firethorn are orange to red. Plant in full sun or light shade, and prune in early spring. A soil pH of 7.0 or so is best for the Pyracantha.

On the north grounds on either side of the north fountain were two old quince trees, which I thought should have been cut down.

However, the flowering quince (*Chaenomeler cydonia*) is a popular early-spring-flowering plant and will attain a height of from three to ten feet. It will bloom in March until April and will even flower in the latter part of February if weather and temperatures permit. The quince bears green-yellow fruit, which is large; the quince is pruned after flowering in spring. Plant in ordinary garden loam, and expose in full sun or light shade. Plant in spring or autumn.

Quince spread by suckers that come up from the shallow growing rooting system. The quince must be suckered if it is to be kept looking presentable.

Fire blight and scale both attack the quince; therefore, control the latter by a dormant spray of lime sulphur and the former by removing the infected parts of the shrub. In treating fire blight, sanitize the cutting shears frequently during branch removal.

Choenomeler lagenaria or common flowering quince is probably what we had on the North Grounds. It was hideous because of its age, and I always had problems cutting suckers.

The forsythia (*Forsythia* species) is a popular border plant, and, although only one or two specimens existed on the grounds of the Executive Mansion, it does deserve mention here. On the South Grounds, several smaller forsythia plants had grown into a mass of dense growth, whereby only a heavy cutting could possibly bring the plant into symmetry.

Forsythia are widely planted early spring-flowering plants and are an easy culture. They flourish in most soil types and can thrive in full sun or partly shaded situations. The forsythia is relatively free from insects and diseases, and its bloom is in demand for indoor floral arrangements. Forsythias do not need a heavy fertilizing, because such doses tend to stimulate foliage and lessen the flower display. Plant these spring beauties in either spring or fall with a five to six foot spacing between plants. Prune the bush after flowering. While on the White

House staff, I was delegated to help prune the forsythia on the lower South Grounds, which had propagated itself to unbelieveable size through layering of branches on the ground. This produced an onslaught of new plants that advanced well beyond the normal edging I had cut around the bush. New plants can be removed by digging and transplanting to a new location. All forsythia upon reaching old age, will develop woody stems which, while not too unsightly, will bear little real foliage or bloom of any value. A program of cutting away old stems should, therefore, commence after the yellow flowers have faded. Several old branches can be removed each season so that the bush does not look too thin at any period of time. I have cut the oldest growth first and proceeded later to prune out other, less favorable growth. Removal of the less vital growth encourages new sprouts, and what was once an old bush can be brought full cycle to a newer specimen.

At the White House are literally dozens of hollies (*Ilex crenata, opaca, convexa,* etc.) which have been incorporated into hedges, walls, clumps, and specimen plants. The large hedges of hollies which formed a protective cover on the lower South Grounds contained many varieties such as *Ilex opaca, Ilex aquifolium, Ilex glabra, Ilex latifolia, Ilex compacta, Ilex rotundifolia* and *Ilex cornuta.* All are evergreen, and, when mixed with deciduous trees and shrubs, as they were at the White House, they form a respectable mass of beautiful foliation. The deciduous mix includes viburnum, magnolia, *Aesculus hippocastanaceae* (Horse Chestnut), *Poncerus trifoliate,* Cornus-dogwood and Acer-maple. Hollies are dioecious, meaning that the berries are borne by the female tree, if the blossoms are fertilized by the non-berry-bearing male. Hollies are best situated in a soil that has been well supplied with peat moss, leaf mold or compost of low acidity; hence the beds for the hollies on the South Grounds. The *Ilex* will sucker as all hollies will, including those on the White House grounds. Sucker hollies during the winter, preferably the latter part of January and February. Pruning for Christmas decorations can be done before the Yuletide.

Exposure of the hollies can be as it was at the White House, shade/sun. Hollies make wonderful understory trees and are rather well adapted for foundation planting. In the latter usage, however, precautions are in order if the foundation is one composed of lime and brick. Hollies need a pH of at least 6.0, and a periodic application of an acid-based fertilizer must be worked into the soil under the trees. Hollies can reach a height of from three to fifty feet, depending on the species. *Ilex convexa* can attain an eight-foot height but can be pruned to a two-foot hedge. It does well with either use although because of the pattern of leaves, *Ilex convexa* serves well as a hedge in a location with full sun. The holly hedges of the East Garden were of the *Ilex convexa* species and were spectacular in their niches of the garden.

The American holly (*Ilex opaca*) is an upright, tall grower and is best used for specimen purposes with similar species in a group planting. The hollies are not by any means immune to disease or insects. The holly-leaf miner can disfigure the foliage of the American holly. Control this pest by applying malathion during spring and early summer. Leaf spot will attack plantings, and, if detected in time, it can be controlled by removing infected leaves. Spraying with fermate or dusting with the same mixture will prevent leaf-spot spores from entering foliage.

Hollies can be propagated by cuttings, seeds, and grafting, I have tried cuttings with reasonable success. Seeds have fallen from some of the hollies on the South Grounds and have sprouted themselves in the bed under the parent trees. I had noticed several holly seedlings in such areas while working on that part of the eighteen acres.

I believe everyone loves the beautiful magnolia, and it is the saucer magnolia (*M. soulangeana*) that is one of the favorite and most widely planted of the large-growing flowering shrubs. The large, saucer-shaped flowers cover the shrub before the foliage begins. With pinching, the saucer will expand into a compact, multi-branched shrub.

Many gardeners without a doubt desire the southern magnolia (*M. grandiflora*) because of its large white blooms in May and June. It is one of the few broadleaf evergreens to be found in the Washington area and may be injured or killed during an abnormally severe winter. I thought we would lose the famed Andrew Jackson magnolia in December of 1968. It lost many leaves, and, as time progressed, the branches in its crown could clearly be seen from the South Fence. The starlings roosting in the Andrew Jackson magnolia with their constant fluttering about helped to dislodge leaves from the tree.

The magnolia can be used as a shrub border, like the Andrew Jackson magnolia and Warren G. Harding trees, and can be used for cut flowers. Although I prefer spring plantings for this as well as other evergreen types, autumn planting will not hurt the magnolia. The magnolia will grow from ten to ninety feet depending on species—*M. grandiflora* will attain the latter number, and *M. soulangeana* the former. Magnolias bloom from late spring to late summer and will bear seed pods with bright-red seeds. Propagate the *M. grandiflora* by snipping spent pods from the branch and extracting seeds. Plant seeds in a peat moss/loam medium in a wooden flat set out in a cold frame. Seedling sprouting may take several months if planted by September; expect seedling generation by the following spring.

Exposure for the magnolia includes full sun for the *M. grandiflora* and light shade for *M. soulangeana*, the latter residing in shady nooks in the East Garden and West Garden, and the former in full sun on either side of the South Portico and gracing the lawn on the Tourist Entrance at East Executive Avenue.

The *Ligustrum* species, better known as California privet by the layman, is one of the hardy plants which we had to maintain on the grounds. The California privet (*Ligustrum ovalifolium*) forms a hedge along the walk west of the South Portico under the Andrew Jackson magnolia. Here it suffers for want of moisture and sun and has never done its best growing during my time at the White House. *Ligustrum ovalifolium* as well as *L. vulgare* and all other species need, at the most, light shade.

Privets are an evergreen group of shrubs used primarily for

hedges, screens, and as background plantings and enframements. This species of shrub can tolerate a wide variety of soils and urban conditions that other plants will not.

Privets can be easily transplanted even without a ball of dirt covering the roots in spring or autumn. If the privet suffers from winter freeze or from transplanting, prune the plants to within a foot or so of the ground, and it will recover quickly.

Most privets produce racemes of white to off-white flowers during the summer, which are followed by blue-black fruit by autumn. The flowers are odious to some people and fragrant to others. *L. lucidium*, as I remember and will never forget, produces a nauseating smell when in flower, and it was all I could do to hold my breath while wheeling tubbed tree specimens onto the back elevator and up to the third floor solarium during the summer of 1970.

Privet does nicely in wooden tubs when the soil medium is properly mixed and liquid nitrogen is applied as a feeding every three to four weeks. Soil for tub privet is abundantly supplied with peat moss and a well-draining soil such as loam. Privets in the tub need to be carefully monitored so that dehydration does not set in and injure the specimen plant. Pinch the soil in the tub to determine moisture content, water if in need, and mulch with Michigan peat moss or bale peat moss to retain tub moisture.

Rhododendron (*Rhododendron species, hybrids*) like a bit of shade, and a few cannot at all withstand full sun. One commonly planted, the Rosebay, cannot tolerate full sun and grows in deep shade and moist soil in the Allegheny Mountains. Catawba and Carolina rhododendrons, however, can tolerate considerable sun. Plant the rhododendron in well-drained soil replete with leaf mold to retain acidity. Never plant in heavy clay soils, where water may tend to stand near roots. The rhododendron has shallow roots like the azalea and, like an azalea, should be planted with at least one-fifth of the root ball on top of the soil. Protect the rhododendron from strong winter winds that may dry the plant and burn the foliage.

We of the White House mulched all of the dozens of rhododendron before the advent of winter to help thwart freezing of fine plant roots. Mulch should remain on the ground in the summer to help cool the soil and maintain moisture equilibrium.

Most of our rhododendrons grew well on the South Grounds. An oak-leaf mulch, which we never used, when scattered through the rhododendron bed, helps to maintain soil acidity as above stated. However, if yellowing leaves appear, supplement the pH of the soil, using sulphur.

The above recommendation should be followed for a healthy bed of rhododendron that will be relatively free of disease and insect infestation.

If care is wanting, then the plants may be attacked by lace bugs, which can be adequately controlled by a malathion dust applied in the late spring and early summer.

Rhododendron are difficult to propagate without controlled greenhouse conditions. Even then, unless cuttings from vigorous wood are taken, the scion may never develop roots in any medium. There is no salvation in using Root-tone, a stimulant, furthermore. While working in the Kenilworth greenhouse, I never was able to root a rhododendron successfully through any conditions whatsoever.

Grafting is possibly the best method of propagating the rhododendron, and there may be merit in layering.

Most of the rhododendron plants on the White House grounds seem to have been there since time began, long before my time and surely long after my departure from the staff. Therefore, I will not be able to record rhododendron species, although I believe such species as *R. racemosum*, a light rose grown in the shade, *R. moximum*—Rosebay—white to pinkish for shade, and *R. minus*—Piedmont—lilac-purple for shade are in the collections by the Southwest and Southeast Gates.

There are several dwarf varieties on the lower South Grounds, about two hundred feet southwest of the East Gate, which I may have recognized as Blue Tit—a bright blue and Impeditum—a purple blue.

Viburnum (Viburnum species and hybrids) comprises a

large and most certainly varied group and includes some indigenous shrubs useful for landscape planting. The leatherleaf is evergreen, and Burkwood semi-evergreen.

Most viburnum require moist soil that is loamy and has a pH of approximately 5.5 to 6.5. Viburnum is planted in autumn and spring, but spring would be indicated for the White House. Blooms occur in March through June, depending on the species, and can be planted in shade and sun.

Aphid deluges are the principal malady of the viburnum and can be removed by soapy water, which is loathed by the insect. Spray also with a tank of malathion; rotation of sprays may be suggested for those aphids that are tough.

Several viburnum were incorporated into the landscape scheme *i.e.*, hollies on the lower South Grounds, and, in particular, *V. sieboldi* and *V. burkwoodii* were in keeping with the evergreens.

Most Acer (Maple) are trees. On the White House grounds, however, the Japanese maples, *Acer palmatum, are shrubs. Acer palmatum does not exceed twelve feet in height, and the leaves of these beauties are maplelike; they have five or seven angular lobes. The lobes in a few of the Acer species are subdivided further into numerous segments and may look more fernlike than what they really are. The various leaf-dividing forms have been named such as dissectum, filicifoluim, linearilobum,* and *septemlobum.* This shows leaf character.

Japanese maples will thrive in full sun or partial shade in varied soil types. Wet winter soil is not favored by the Japanese maple, so a raised planting is suggested. Pruning of the maples is not necessary, but winterkill of branches from cold air and frost must be cut out in the spring.

Propagation is usually by grafting onto seedlings of *A. palmatum*, which is done in the spring in a greenhouse.

Acuba japonica or spotted laurel did grace the South Grounds near the Southeast Gate and was a large plant in that location. The acuba is an evergreen with large shining leaves, either light green or green spotted and blotched with yellow. The latter variety grew on the White House grounds. The variegated form is called spotted laurel. The acuba will grow eight to nine feet high, is a bushy shrub and is unisexual, *i.e.*, some specimens produce only male flowers, while others produce only female flowers. The females give fine crops of red berries in winter if a nearby male bush pollinates.

The acuba will grow in most any soil and can be planted in the shade and under larger trees nearby. It was this situation that existed for the acuba on the South Grounds.

The acuba does not need to be pruned, but can tolerate pruning if so desired. I never pruned the acuba while on the staff. If pruning is desired, however, prune in April.

The common acuba grows from seed sown in April in a greenhouse. Seeds vary, so selected specimens can be increased by cuttings taken in summer from firm young growth and rooted under a fine mist of water, or ripe cuttings taken in the fall and rooted in an unheated cold frame.

Cornus (Dogwood), three of which graced the grass triangle near the Southwest Gate (*Cornus florida*), are trees and will be treated here along with the shrubs. Although the *C. florida* grows to a fifteen-foot height or more, those on the South Grounds were young and small at the time of my entrance onto the grounds. The *C. florida* serves as a good understory tree, and its flowers add life to the spring garden scheme.

Elaeagnus (Oleaster) is one of the more prominant shrubs on the grounds. It was planted around guard booths and other utilitarian structures to help blend these man-made obstacles into the landscape. The species seen on the grounds was the *E. angustifolis* with narrow leaves, silver undersides and gray-green tops. All will grow in "any ordinary soil" and full or half-sun. The quotes are added, and the "ordinary soil" is surely true, even to the point that the shrub will grow on top of concrete. *E. angustifolia* were used to hedge around the Halderman patio. *E. angustifolia* and the other species do not need to be pruned, but do need to be suckered once or twice per year. Suckers detract from the beauty of the shrub and cause an apical dominance to occur within the bush.

Propagate *E. angustifolia* or any of the other species by

cuttings of firm young growth in the summer in a cold frame or under a sporadic fine mist of water in a greenhouse.

Euonymus (Spindle Tree) is comprised of both evergreen and deciduous species and each served different needs in the garden. The form most common on the White House grounds was the low creeper groundcover *E. radicans*. This is used primarily on the East Executive Avenue side of the grounds on the hillsides and not only covers the hill, but freed us from having to cut grass at these spots. The *E. radicans* is a good cover to help prevent erosion. Formed by successive layering of lead vines, *E. radicans* will overshoot its confines and layer into the lawn. This activity can be halted by removal of the layer stem and by periodic trimming of the *E. radicans* bed. Trimming will also prevent *E. radicans* from working its way onto and up nearby trees and shrubs. If allowed to climb trees, it will act as a parasite, filching needed moisture and nutrients from the host and eventually killing the tree.

E. radicans is easily propagated in sand, perlite or a mixture of palite and Michigan peat moss. Allow two to three weeks for rooting to begin and adjust before removal from the propagation bed and subsequent planting in peat pots. Cuttings will also root in a jar of water and can be raised hydroponically.

Osmanthus is of the olive family and is an evergreen shrub with small flowers which are white and redolent to the nose. This shrub is often seen in English gardens, but it is also seen flanking both sides of the South Portico of the White House, enclosing the East and West Gardens and growing in the East Alley flanking a granite staircase. *O. ilicifolius* is the most popular species and is an osmanthus. It is a large shrub and looks like holly in leaf and growth. It may attain a height of fifteen to twenty feet if left on its own. It is best grown as a hedge.

The osmanthus will produce white flowers in the early fall and has many garden varieties such as *purpurascens*—purple leaves and *aureomarginatus*—leaves variegated.

The osmanthus will grow in any good soil and can be planted in full sun or partial shade. A heavy shade tends to thin the foliage of the osmanthus. One does not need to prune the osmanthus, but it does well as a trimmed hedge. Propagation is by cuttings of firm young growth in late summer in a propagating frame or under a fine mist of water in a warm soil.

To bring attention to the East Garden fish pool a *Poncirus trifoliate* (Hardy Orange) graced the patch of ground on the left of the pool facing east. This is a deciduous shrub, stiff and angular in branching, and grows up to ten feet high. It is fortified with spines, which are painful to the touch. The tree has fragrant white flowers in late spring, followed later by orangelike fruit. *Poncirus trifoliate* likes full sun, but can serve as an understory tree in partial shade. It will survive in a fairly good soil and can be propagated by seed sown in a warm greenhouse in the spring. *Poncirus trifoliate* will also self-seed if its fruit is not raked and disposed of after it drops in the autumn. The cluster of hardy orange near the tourist offices was usually left to itself, and, consequently, the oranges rotted on the ground while some of the seeds germinated to form young trees.

Poncirus trifoliate will also, like many shrubs, sucker if not watched closely. Suckers reaching five feet have been removed from those specimens near the tourist entrance.

The West Garden is also known as the Rose Garden, which may seem ironic since roses are hardly the rule, but surely the exception among the plants found within the garden. While it was under my care, there were a dozen small rose shrubs which were usually eclipsed by other annual and perennial plants.

Roses require full sun and as little as half-sun may have fallen on those roses in the West Garden. This, coupled with a frequent cycle of deep cultivation and heavy watering, caused many of the roses to simply turn brown and die. Replacements were brought in as needed, but I don't believe they ever served any purpose other than as an occasional accent to some nook or cranny of the garden.

Wisteria graced the wrought iron railings on the steps leading up to the first-floor South Portico porch. The most common variety was the *Wisteria sinensis* or Chinese wisteria. This is a very vivacious climber and can easily reach twenty to forty feet with densely packed flower tracts up to one foot long.

The Japanese wisteria, or *Wisteria floribunda,* is not as vigorous as the *W. sinensis* and is variable in its flower trails. Some types may have flower trails not much longer than those of the *W. sinensis.*

All wisteria are deciduous and can be raised in rich soil. They are sun-lovers and do well on south and west walls. They can be trained into shrubs and standards with large crowns on a fairly straight stalk. Do this by pruning each midsummer, when the young shoots are pinched back five or six inches.

Wisteria are best generated by layering in the spring or summer. The boxwood (*Buxus sempervirens*) is considered one of the more choice of garden plantings, and the English Boxwood is especially revered for its grace and solidity. The boxwood has been cultivated in the east since colonial days, and its rich green foliage without a doubt increases its popularity. This plant is relatively free from pests and can tolerate varied soil conditions.

The boxwoods at the White House were used as the main feature of landscape on the North Grounds by the portico.

Today, botanists classify both the English and American boxwood as the same species. Foliage and growth-habit variations are given clone names. The *B. arborescens* (American) is one type that is referred to as American Boxwood. *Rotundifolia* is the variety having broad oval leaves and may be the hardiest of all boxwoods. Other varieties include *B. sempervirens, var. bulata* and *angustifolia.*

Besides the English Boxwood, there is one from Japan, Buxus micraphylla. This species is represented by japonica and koreana. The former was used as a low hedgerow along the lawn in the East Garden.

Boxwoods require shallow planting, and, if planted too deeply, they will suffer. This, together with close cultivation, tended to harm the English Boxwood in the West Garden. Boxwoods have shallow rooting systems, and extreme caution must be exercised when cultivating around them. Planting flowers, whether they be tulips or balled annuals, near the boxwood will surely displace too much soil during planting. More often than not, the novice will shove the extra soil up against the boxwood, causing it to smother.

Boxwood should be planted in the spring, and at this time the gardener ought to add compost or well-seasoned cow manure in generous amounts. Never fertilize the newly planted *Buxus* with chemical additives. When the *Buxus* is established, excessive feeding should be avoided, as the boxwood can be fed in alternate years.

Boxwood can suffer from infestations of the leafminer and red spider mites. The mites thrive on the underside of the boxwood leaves, making control harder than usual. A program of spraying with malathion or carbaryl will help control these pests.

One other problem that confronted me on the White House grounds concerning the boxwood in the Harry S. Truman planting in particular was "die back." It is caused by the collection of dead leaves in shrub crotches. Nectria is found in vegetable decay and is the cause. Nectria starts in the clumps of dead leaves, and its roots pass through the mass of leaves into the moist bark of the surrounding branches. The leaves then become small and pale as the roots of the Nectria permeate the branches. If not tended to, the affected branches die. The best method of control, and the one I used, was to hand-pick the crotches and remove the dead-leaf masses. Too, one can elect to use a hose for clean-up operations or a small broom will suffice for the job.

Although we did not protect our boxwood at the White House for winter other than with mulch, it is advisable to do so by using slat coverings. Protection of this sort will help to maintain the boxwood shape, should a destructive, wet snow fall. Boxwood hedges where the soil may be looser and softer than some other areas may require burlap covering or pine-bark mulching. In the Rose Garden, I prepared my boxwood hedges for winter by using peat moss by the bale and dressing around the plants. This treatment would keep the soil in the immediate sector reasonably stable during the freeze/thaw cycle, protect the shallow roots during warmer weather, and help cut out broad-leaf weeds and the germination of *phreatophytes* brought in on wind currents.

In the East Garden, where my boxwoods were trimmed into hedges for the most part, little mulch was used. The garden was protected by its location—close to the East Colonnade and tourist lobby—and by being situated on the south side of the mansion. The green pillow boxwoods with their nodular profiles on either side of the East Garden lawn, running east to west, would shed leaves, providing a natural mulch throughout winter. It is, however, more healthy for these expensive shrubs, if such dross is removed during warmer months. This was done after each trimming in the summer.

Most of my experience with raising the yew (*Taxus*) had come while working with the Capital Parks crews in 1965, and, by the time I reached the White House, the yew was rarely seen, aside from a few plantings near guard boxes, one gigantic specimen on the South Grounds, and the large, rounded specimens along the concrete walks flanking the North Portico. The *Taxus*, as a rule is a sun-lover and tolerates almost any reasonable soil. A loam is best, but the yew will grow in a clay medium as well.

Most evergreens, including the yew, do not benefit by assiduous stirring of the soil, but will do well with a generous helping of mulch. For good results at the time of planting, a soil that is replete with humus or peat moss is needed. This application will conserve moisture for the roots, which is important for their health. A dry conifer root will never revive, and the gardener can consider the plant a loss. Evergreens need a good supply of moisture in fall and early winter.

Evergreens do not require lime as they do best in an acid soil. If mulch is heavy around the plants, then fertilizer is seldom needed.

Yews need an annual pruning to keep them compact like those on the North Grounds. Trim in March or in September before the new growth commences. If left to themselves, yews will lose their compact figure and will spread their branches outward. This unkempt growth may be desirable as varied plant profiles spice up a landscape scheme.

Yews such as the *Taxus cuspidata* can be planted in sun or shade and will thrive in both environments if the soil around them is kept fairly moist.

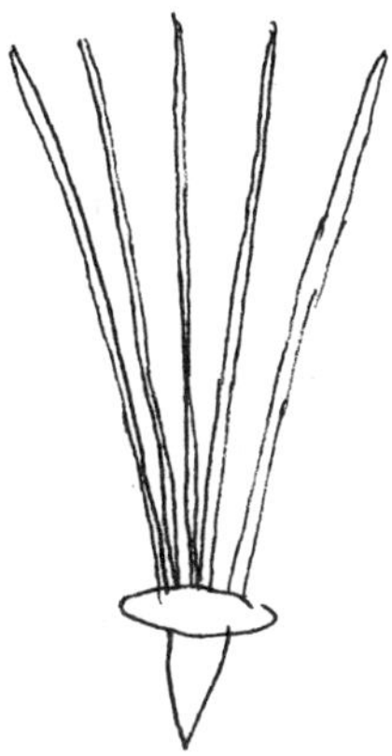

Pinus strobus

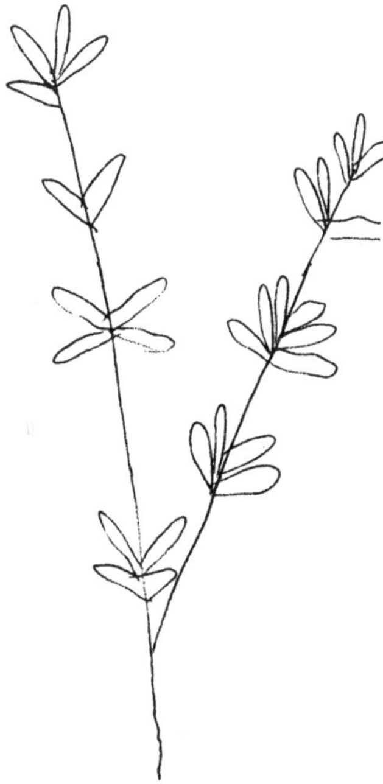

Rosemary

White House Herbs

Many of today's gardeners delight in the cultivation of herbs and in the field of growing herbs. However, not all cooks make use of all herbs. Only a few kinds are in frequent use, and these include rosemary, Saint Helenia and thyme.

Few herbs should be raised in the beginning until the gardener becomes used to them. There is no sense in growing herbs unless they are to be used in some manner, *i. e.*, in the landscape or in the cooking pot. Raising herbs is not as difficult as one might expect, and they do not need too much space in the garden. Herbs should be relegated to one side of the garden or in a carefully set-up garden scheme such as that of the East Garden of the White House.

Most herbs can be grown in any rich, well-drained soil, which means a program of special soil preparation. Using manure or compost, work the garden to a depth of about one foot. To increase soil moisture-holding capacity, add bone meal or cottonseed meal. If the garden is of a clay-soil type, add perlite or sand to lighten and aerate the soil. The deeper the garden extends with the above preparations, the better the garden will be for herbs that will be planted.

Chives (*Allium schaenoprasum*) are very popular as perennial plants and are as easy to raise as onions. Once begun in the garden, they will last for many years. Those chives in the East Garden were quite vigorous and rarely needed to be replaced. Chive seeds can be sown in the garden if so desired, but ours were brought in as potted specimens and transferred directly to the garden as needed. The leaves of chives are used in flavoring various foods like meats and salads. Our chives in the East Garden were allowed to reach only six to eight inches high before being cut. The size was then three inches. Chives, if left to themselves, will quickly become weeds. A bi-monthly trim is therefore a must. Chives transplant easily and will grow well in garden loam. This herb takes a position under full sun for maximum growth. Chives should not attain a height over one foot.

Sweet marjoram, (*Marjorana hortensis*) also used in our landscape layout in the East Garden, is a perennial herb that requires full sun for best growing results and a light medium such as a rich soil plus average moisture. This herb can be raised from seed, and the germination process is slow to come about. Sweet marjoram will grow up to one foot, and, when harvesting, use it as it is.

Sweet marjoram is used fresh for seasoning meats, soups,

and as a dressing. It can be grown as a pot specimen, which I did for potted plants in the two gardens, and can be potted for house use. This herb can also be dried and its essence used later. If dried, sweet marjoram is needed; then cut its tips before the plant flowers.

Another herb to be used in the East Garden as in the greenhouse on the roof of the White House is the rosemary (*Rosmarinus*) herb. Rosemary requires the same conditions as stated for the other herbs. Rosemary attains a height of around one foot and will give off a most fragrant odor when cut or rubbed with the hand. It is a fine herb additive for cooking, as our chefs know. I have often cut a few pieces of rosemary to take home for garnishing a steak.

Thyme (*Thymus*) requiring the same gardening techniques as the rosemary herb and sweet marjoram, is also found in the East Garden. It requires full sun and may become scraggly if left in the shade or if it is not pruned on schedule. Thyme makes a fine specimen plant if in full sun. Our thyme was cut into a small square hedge under the topiary holly in the East Garden.

Saint Helenia, another of our fine White House herbs, is also a perennial and was much in demand as winter herb hedges and borders in both of the Presidential Gardens. Saint Helenia was also to be found as borders at the North and South Fountains in some winters. Saint Helenia can be brought in either the silver-gray or emerald-green, and the former color was used in the West Garden as winter hedges under the flowering crabapples. A diamond-shape diagram using this herb was set up to follow the angularity of the Buxus hedge, also diamond shaped on two sides.

Saint Helenia works well as a hedge if initially the Saint Helenia has a mass of growth from which to trim a bevel-hedge profile. This herb can be transplanted in the spring, and, if you transplant in the autumn, I would recommend a ball of earth to accompany each plant. This was done when this herb was brought in for the winter display. When first received, the Saint

Helenia were a mass of impossible growth, but after carefully planting I followed with my shears to cut the bevel and form the hedge.

When using Saint Helenia for regular borders as we did at the North and South Fountains and in the small garden by the South Portico, the plant does not need to be trimmed too fashionably, but can be left with its rounded head of silvery growth.

Saint Helenia can be treated much like the other herbs except that too much watering may rot the roots of the plants and cause the silver-gray or emerald-green growth to droop and look unsightly. Trim sparingly or not at all during hot weather, as the shock of pruning can cause the Saint Helenia herb to die. The Saint Helenia can be pinched and used in cooking if so desired.

Weeds of the White House Lawn

Bermuda Grass—*Cynodon dactylon*
Ground Ivy—*Glechoma bederacea*
Clover—*Trifolium*
Chickweed—*Stellaris media*
Dandelion—*Taraxacum officinale*
Crabgrass—Foxtails—*Alopecurus*
Onion—*Allium schoenoprasum*

The list above contains some of the weeds I was familiar with while working on the White House lawn. The common dandelion, which is very hardy, was one of the best growers among the weeds, and I spent many days trying to control the weed. Others such as the *Allium schoenoprasum* probably got started on the South Lawn from wind-borne seeds originating in the East Garden or from origins elsewhere. *Stellaris media* was common in some parts of the lower south grounds but was easily recognized and disposed of when seen. Ground ivy or *Glechoma bederacea* was another hardy weed which, if not too extensive, could be ripped out by hand. There were not too many occasions of this lawn pest being discovered on the grounds. *Trifolium* was quite common on the North Grounds in areas where resodding had not been done for several years.

Another weed not mentioned above was the nut grass that infested parts of the lower south grounds, especially around the South Fountain. It was easily seen because its blades towered over the K31. The nut grass also had a lighter shade of green. Any good weed-killer was used for small patches of weeds on the Executive Grounds, but extensive infestations were dealt with by cutting out the infected sod and importing new sod. Several acres of lawn were cut up and discarded in 1969 because weeds were manifesting on this acreage. It was easier to take up the old sod and relay with new. A family of four could have lived in opulence with the money spent on the White House lawn.

Other White House weeds included: morning glory, lambsquarters, pigweed, prickly lettuce, ragweed and vetch.

The second from the top, lambsquarters, was less common on the White House grounds than in Washington at large. This weed is edible when young and can be prepared much like spinach or pokeberry. The mature lambsquarters is tough and ugly and will grow to six feet or more. This weed was occasionally spotted growing near garden water hydrants and was pulled accordingly.

Turf builder would help to control weeds as this chemistry encourages the rooting of desirable grass which in turn will smother out the broad leaf species. Weeds will grow if turf becomes weak-rooted and spotty.

White House Greenhouse
1972–73

Even with the multitude of flowers that had to be cared for on the Executive Grounds, such as the Rose Garden, Jacqueline Kennedy Garden, the two fountains and the entire acreage, and so on, the White House had a rooftop greenhouse that also required pampering. Not until returning to the White House in

Grape

Pansy

1972 did I become fully acquainted with this rooftop flower collection. I had known of it before. In the latter 1960s other people on the staff were boarding the back elevator and watering the flowers on alternate days. The greenhouse was small, measuring seventeen by eight feet and quite fully complemented with flowers at certain times of the year. These plants were the personal property of the First Family, although some belonged to the flower shop, some to the house staff while other flowers and herbs were used by the White House VIPs and kitchen staff. This little house of glass also contained a propagation bed with heating units and an automatic misting system for moisturizing the almost eternal rows of resident deciduous and evergreen cuttings. Some of the flowers I helped propagate here "on the top of the world" as we called it, included the border plant dusty miller, a few rhododendron cuttings, *Taxus*, geraniums, English Ivy and coleus, the latter of which is an herb.

The primary duties attached to this greenhouse entailed among other things, watering, fertilizing, and collecting potted house plants and delivering them to this sunny roof garden to be revitalized. Later, these plants would once again grace the various locations downstairs in the mansion. I was especially honored to have the privilege of caring for Mrs. Nixon's special plant, *Streptocarpus*, which she had delivered to the greenhouse to be revived for several weeks. It had become ill in the Presidential Apartments below.

When I entered the greenhouse on warm summer days, my nostrils were assaulted by the "antiseptic" or scented-leaf variety of geranium. It was kept here for flavor for state dinners when its leaves were harvested for table use and for its scent. The scented geranium was peculiar to the greenhouse and was never seen or used in the garden schemes on the grounds below. The rather definite smell reminded me of a regiment of white sanitized nurses marching through spick-and-span hospital corridors dispensing their sundry nostrums. I wondered if the scented leaves of this geranium could be dried and concentrated for its aroma and used as a cooking herb. I never tried it.

Another plant item of interest found in the rooftop greenhouse was the coleus-in-the-basket, which enhances the usefulness and beauty of this herb. I had two such basket arrangements hanging from a water pipe overhead which added décor and variety to the interior of the hothouse. The method, à la White House, of making this arrangement begins with a wire basket and several handfulls of sphagnum moss to act as a liner on the sides of the basket. Line sides and bottom with sphagnum, then fill the basket with a premixed soil concoction of one-third peat moss, one-third garden soil (loam) and one-third perlite. Fill to within one inch of top of basket, add up to one half dozen small coleus plants. Water generously and set up to hang. Baskets of this sort with sphagnum lining will last over a year, whereupon the acidity of the moss will corrode the steel wire. A new basket will then be needed.

The only place that *Nephrolepis bostoniensis*, or Boston fern, was seen on the eighteen acres was in the greenhouse. This fern variety, together with *Adiantrum*, or maidenhair fern, was kept in the greenhouse for several months at one end of the gravel tray so it could refoliate. The ferns had suffered dehydration somewhere in the mansion below. These ferns do best in a shady environment with an acid leafmold soil as a growing medium. The Boston fern will become quite large, over five feet in diameter, at which time it can be pinched back or some of its branches trimmed free of the plant. My specimen reached two feet in diameter before its owner finally retrieved it from the greenhouse.

Boston ferns transplant easily if a ball of dirt is attached to the root system. The new plant site should approximate that of the old so that the shock of being dug up and relocated is less. Potted ferns may become potbound, at which time repotting to a larger clay container is desirable. Transplanting to the outdoors is also possible.

Another plant under my care was the strawberry plant belonging to the First Family. It too had become in need of better growing conditions and was placed in the hothouse to recuperate. This particular strawberry plant was planted in a white glazed ceramic "pouched" pot, containing perhaps eight to ten pouches from which the strawberry foliage could peek out around the sides of the planter. This is an interesting planting arrangement for a specimen such as the strawberry plant with its never-ending runners and subsequent foliation. This planter was kept in a sunny corner of the greenhouse until one of the First Family came up to call for the plant.

Petunia baskets with English ivy were another variety in the greenhouse, and their method of manufacture was similar to that of the coleus. Again, sphagnum moss was used as a basket liner and loam used to fill the basket to its brim. Therein, plant young petunias with English ivy hanging down over the basket edges. Coleus-in-the-basket need to be pinched to retard flower formation, and so also does the petunia when it begins to flop over the side of the container. In this case, the petunia flowers may tend to dwarf as the plant ages and becomes scraggly. Therefore, one must pinch the petunia back even at the expense of some flower buds and bring the petunia back into its rightful dimension in the basket. Petunias will rebud and flower, however, after a pinching operation. Finger-pinching is in line, or the gardener can use snips or scissors.

The White House greenhouse served as an ancillary source for the needs of the White House flower shop far below on the ground floor under the North Portico. I say ancillary, but few flowers were probably ever taken from this greenhouse. Bulk loads of floral varieties were grown at Kenilworth expressly for the White House. On rare occasions, the flower shop has resorted to nearby commercial flower shops in downtown Washington.

I was quite fond of the *Solanum pseudo-capsicum*, or Jerusalem cherry plants, that I had on reserve in the greenhouse on the "top of the world." These are hardy plants easily raised from seeds of the fruit. In fact, a rotting cherry was seen to produce several small seedlings in a secluded area under the propagation bed one day in the spring of 1973. I noticed them while hosing down the tile floor of the hothouse. Hardly a place for small fry of this Nightshade Family to be cultivating themselves! I had to remove them while sanitizing. The mature Jerusalem cherry plant looks like a miniature tree, considering its branching habit. Such an arrangement of branches may even allow the plant to be used in bonsai settings. I write "may," because bonsai culture usually requires conifer miniatures for microgarden landscapes.

Rosemary and thyme were also potted and raised in the White House greenhouse and were the foremost herbs in the hothouse. The greenhouse being built and situated where it was, on the northwest corner of the roof, was opportune for the White House chef and his cuisine, since all the chef had to do was to ride the elevator to the roof and collect the seasoning for the daily menu. The greenhouse herbs were held for winter, while those growing in the East Garden served the summer menu.

Other horticulture delectables included in the greenhouse repertoire were Christmas cactus, an unusual plant blooming near year's end; several Hosta specimens, which were retained for a later position in one of the Rose Garden borders in 1969, but which never made the transfer; one dumbcane of fairly reasonable size, which I could have used on my veranda at home; several potted dusty miller, the vanities of someone downstairs, and too few for bordering with the geraniums of annual selection in any of our grounds displays; one beautiful fuchsia, which belonged to one of the Nixon girls and which received the utmost care; one recently overturned but retubbed small *Ilex convexa* placed in the hothouse for rejuvenation; four *Heliotropium peruvianum* (the common heliotrope) left-overs from a border in the Rose Garden (1972); and *Cineraria grandiflora,* the private little enterprise of the White House flower shop. The latter plants are delightful for spring place settings on White House tables.

Crataegus (Hawthorn)

62

Tub Plants and Flowerboxes

To continue with the White House horticultural ensemble, I turn next to the various tubbed plants and flowerbox arrangements seen and unseen around the White House grounds, depending on whether the display is seen by the public or simply by our staff who were creating the arrangements.

Probably one of the more graceful trees to be used in the large boxes are the *Ficus*, or fig, in common parlance. These semitropical to tropical trees are used to decorate the East Wing Lobby and are on display for the thousands of tourists to view each year. Therefore, we had to keep these specimens in top shape by pruning when in need, watering the tree boxes and maintaining the base of the tree, *i. e.*, removing dead leaves, light trash, and cigarette butts and dressing with peat moss. These tree boxes were marvelous catchalls for tourist trash. *Ficus* that are tubbed in such a manner need occasional liquid nitrogen, and, when time arose, droopy plants were removed from the lobby and fresh trees brought in. This was the cycle: one month in and one month out to Kenilworth to be refurbished, then returned to the White House for another month. We had several styles of cut for our large *Ficus*—the standard with a round topiary cut at the crown of the tree—and the free style, whereby the crown was unkept in a manner of speaking, but did retain some profile. Both of these styles in the nine foot *Ficus* were used interchangeably in the East Wing Lobby, the East Wing Colonnade and the West Foyer.

Also, in the line of *Ficus* trees we had the smaller, six-foot specimens, which were of definite style—a semi-topiary cut with rather compact foliage. These figs had foliage from the top of their fourteen by fourteen inch boxes to their crown, and the profile resembled a stretched oval or a linear hotdog on end. These trees were kept trim at Kenilworth and, like the larger, nine-foot trees, were cared for in the same manner. However, occasionally the six-foot *Ficus* would miss their watering, and I thought we were going to lose the two of them before they were taken back to Kenilworth for treatment and new trees restored to the East Wing Colonnade. The *Ficus* were strictly our responsibility and not the house staff's. The latter statement is surely the truth, and the domestic staff could not have cared less whether the poor fig survived the heat and neglect of the house. I heard rumor that one of the staff in the house had kindly urinated on one six-foot fig tree because he felt for the suffering tree. It had not been watered in days. Thanks, but no thanks. I should see to the fertilizing of the fig if they pleased.

The *Ficus* inventory included about a dozen of the nine-foot trees—half-topiary and half-free-branching—and approximately fifteen of the six-foot trees, the former in twenty-four- by twenty-four-inch aqua-white painted wooden tubs, the latter in a small size of the square tub.

The *Ficus* added just the right touch to the ritzy interior of the lobby, and they finished off the East Wing Colonnade with appropriate accent needed for that part of the mansion.

Ficus desire up to full sun, and those that are potted can stand some shade for a period of the day. The West Foyer was a little too shady at times, and we had a problem with the larger *Ficus* dropping their leaves before their scheduled departure for Kenilworth. Watering was done at a minimum in the foyer, and to facilitate less cleanup of fallen leaves, I shook the figs in question until all of the leaves fell to the floor. One sweep-up was enough for a day.

While describing the potted trees for the East and West Wings of the White House, I will tell something of the poodle junipers that were placed in the exterior niches on either side of the East Front Entrance and on either side of the marble staircase leading to the Oval Office and Cabinet Room. The first location was not the best for the sun-loving juniper, poodled or not, and I deferred watering these two trees after noontime since the sun did not bathe these trees with warm rays at that time. On hot summer mornings, I would water these two plants first so that the trees could use as much of the moisture as possible before evening. One watering in twenty-four hours was usually enough to sustain the trees. Severe dry spells dictated that I give one or two additional splashes to these trees before the end of the day. It is preferable to water trees in the evening, but a potted tree in the shade may be better off being watered in the morning. This discussion leads up to the fact that the moisture in living plants must be supplied and used so as to benefit the plant. Surpluses and deficiencies will not help the tree to prosper. Because of the location of the two junipers in the east front niches, growth was almost at a standstill. Keeping the poodle profile in check was easy, as no branching ever took place. In essence, I had to rotate the two trees to find the best face to present to the daily tourists.

The two poodle junipers near the Oval Office were placed in a better location where they received much more sun and likewise grew more fully and could be considered poodles. These two junipers, each three feet high, added zest to the Rose Garden and managed to fill out a barren, sun-reflecting spot as these marble steps were. The poodles were off-limits to anyone other than myself and gardener George Hundsaker, in that we did the trimming, fertilizing, and general upkeep.

In addition to the juniper poodles in the West Garden, we also had tub *Ligustrum* set up against the exterior walls of the Oval Office and Cabinet Room—another bad spot for sun-relishing species, but the *Ligustrum* were removed on schedule as were the *Ficus*. During presidential absenteeism was the best time to manicure the plants near his office—water, aerate, prune and fertilize—according to individual plant needs. During extended in-office tenures, it was almost impossible to come within fifty feet of the West Wing complex. I was often shooed away by the Secret Service and could never draw near the area to revive my plants with a refreshing drink of water. As a reward for my unintended negligence, I often found drooped plants in dire need of watering and pruning of dead foliage. This was true also of geranium and petunia boxes, which graced the terrace just outside the Oval Office. We were one up on this; nevertheless, we prepared a special moisture retaining soil medium to help conserve moisture in the boxes that could withstand long bouts of warm weather and the noncaring office VIPs.

The flowerboxes, of which we had four on the Oval Office terrace, by the Halderman patio, together with two tubbed *Ilex* and over a half-dozen smaller boxes in the windowsills of the VIP offices, were especially nice for annual plantings. Since there was a superfluidity of spring bulbs in the Rose Garden, flowerbox decoration was held over until the annuals arrived in late May. The most popular annual selection for the above-located boxes in 1972–73 were petunias and red geraniums, the former draping over the edge of the planter with their burst of white blooms, and the latter more upright with their luscious red flowers. What a color combination in all its splendor! Mums were our favorite autumn flower for these boxes and Bronze Song supplied vigor and life to the terrace.

The annual flower planters located on the exterior windowsills of the VIP offices near West Executive Avenue probably received the least care of all tub plants here and after mentioned. This was so because of the general ignorance of the VIPs about watering the box and pinching the petunia blooms as they died and also due to the fact that the boxes were occasionally forgotten by our staff. These sills were hidden from view in all directions by a row of seven-foot hemlock and I never understood why we even bothered with these planters. We went in behind the hemlock hedge one day and removed all boxes. All were dry and the flowers brown. We did not replace them.

We now move to the third-floor solarium, which faces the Washington Monument, and Jefferson Memorial. The roof of the White House had a tile walk that ran around the top of the mansion, excluding parts of the North Portico delta roof. The solarium was for the use of the First Family and included within an octagonal room of ten to fourteen feet, a television set, a stereo, a small library, the usual accoutrements for relaxation, and other amenities. Outside, and facing the solarium was where we placed an assortment of tub shrubs and trees. The trees were *Pinus strobus* (White Pine), ten or more in number and all within six feet in height, all residing in eighteen-inch, dark-green wooden tubs. Next came the *Ilex convexa* in the above-size tubs and all two and a half feet in diameter. At least one half dozen of the *Ilex convexa* interspaced with the *Pinus strobus*, creating a portable hedge and rendering the exterior of the solarium more woodsy. The solarium had glass panels and—you guessed it—they were bulletproof. One might wonder if the First Family feels hemmed in by so many closely spaced trees and shrubs.

Between the tile catwalk on the roof and the balustrade at roof's edge was a wall where we flanked the evergreens with tubs of red geraniums and petunias, which gave color to the otherwise verdant cover.

The solarium was a most important focus for the Nixons, and it was always imperative to be mindful of the shrubs and flowers we had installed in the vicinity of this sun room. I was primarily responsible for the tendering of the floral decoration—watering, pinching, replacing spent plants and fertilizing on time. All of these plants were recycled to and from Kenilworth during the same trip that covered the other potted plants at other locations in the mansion. I often thought how well the potted plants behaved considering how frequently they were moved around and the subsequent jarring of their fine roots. I did lose one white pine when it fell out of its tub during rough handling.

Nevertheless, evergreens will grow well in a tub environment if the correct soil is chosen (well-drained, peaty soil) and with this medium being packed well so that roots do not dry out. The tub culture requires close observation—watering, fertilizing, and pruning when needed—will keep the tub plant growing its best. Fertilize about every five to six weeks using liquid nitrogen. Peat moss mulch can be used to dress the top of each tub to conserve moisture and keep the tub soil cool.

We now peek over the South Portico roof and focus our attention on the two eight-foot, boxed *Robinia pseudoacacia* (Locus) sitting on the marble pavement on either side of the South Portico canopy. The *pseudoacacia* do lend themselves nicely to tubbing, and adhering to the conditions outlined above for the evergreens will serve well for ornamental trees. Our two specimens were topiary, or standard, and drew attention to the ornate entrance area of the South Portico. The *pseudoacacia* may be fussy with water requirements when it is planted in a tub. Our two locusts gave off as much as ten gallons of water per day per tree, and frequent moisture recharge was a must if the foliage was to look its fullest. There is nothing worse than a water-starved deciduous tree with its leaves fading for lack of moisture. Locusts have a shallow rooting system, and this must be taken into account when potting. Tub-bound specimens are not uncommon, and both a crown and root pruning may be needed after the tree has been grown in the tub for a year or more.

We had several of the *Ilex Americana,* which reposed on the roof over the West Wing and which were neglected more often than not. Nevertheless, these tough trees did live on for the time that I was on the staff. These trees served as backdrop during various outdoor functions on this roof and were never moved from the spot. This was fine, as each tub was about four by four feet with a half-ton of soil in each of the containers. They were therefore, rarely moved from one place to another. It was remarkable how the roots of these *Ilex* never broke free of the tubs to search out water and nutrients.

The next flower-bedecked point of interest is the stone footing that supports the four massive Ionic columns of the North Portico. Until the early 1970s this area was drab—nothing but the boring, eternal white painted bastion reflecting the summer sun. In the beginning of the 70s, this area under and between the columns was decorated with large wooden flowerboxes made to order by the carpenter shop to fit neatly between the four columns. These boxes were six by two feet and were compartmentalized so as to add strength and durability to the box and to segregate certain flowers from other species; planting was easier with compartments. The old favorites, red geraniums and snow-white petunias, were the summer occupants of the three large boxes; autumn displays included the lighter-colored mums, such as the yellow, bronze, and white.

These boxes added to our weekly workload in that the annuals, in particular, had to be pinched back (petunias), watered and checked for parasites such as the aphid. Furthermore, the geraniums, being vigorous and a healthy species in the prenourished soil, would bunch up and begin to choke one another out. We, therefore, had to aerate the plants by plucking oversized leaves free of the main plant stem. This left unsightly yellow-green leaves exposed to the watchful eye of the tourist and the TV camera. Stale leaves of this coloration were also picked. The plants were once again left to grow into another entanglement. Since petunias like to "run away" with vivacious growth, prudence was expended in trying to keep them contained within the limits of the boxes and not allowing them to fall all over the exterior sides of the planter.

The annual flowers used in the above location were just that—annual—in terms of a long-term insecticide dispatching program for the eradication of pests. These flowers were not too readily affected by aphids, white flies, or spital bugs, so that spraying was held to a minimum. Again, being annual, the geraniums and petunias were hurriedly pulled and dumped upon the arrival of the autumn arrangements. Thus, the annual summer flowers might grow in the planters for a dozen weeks. If we had no floral turnover; geraniums and petunias could well last until the first killing frost.

Planter-media stratification, as far as plant growing was concerned, included the following concoction: The bottoms of the flowerboxes were liberally packed with pea gravel to afford drainage. This gravel pack might be up to three inches. Over the gravel, a perlite-loam mixture was added as the soil for the geraniums and petunias. After planting the annuals, the boxes were watered and left to produce their blooms.

Last and surely never least, if the VIPs had their way, was the flowerbox display situated by the VIP entrance at the end of the West Alley. This entrance, serving the staff of the President, the news reporters and others of titles and letters, resembles a small North Portico. Columns rise from a small footing, and a roof protects chauffered personnel during inclement weather. Flowerboxes placed at this location are a repeat of that arrangement seen at the North Portico. The boxes are smaller, but contain the same flower types. The flowerboxes are complemented by adjacent clumps of azaleas, *Ilex* and K31 turf. The Petit Portico is well-landscaped.

This then, in essence, encompasses all of the potted and

tubbed floral decorations and arrangements that beautified the many points of interest in and around the Executive Residence, from the East Lobby to the Oval Office; from the "top of the world" to the South Portico; those flowers of the North Portico to those seen and cared for at the VIP entrance. May these shrubs and flowers all grow and bloom in splendor, encore, and encore, and encore.

Ilex opaca

The White House Flower Shop

Another facet of White House horticulture and perhaps one of the more interesting and challenging aspects thereof was the flower shop, first located in the West Wing complex, housing the Presidential swimming pool, but later moved east to a larger accommodation under the North Portico. The latter location was more suitable in that all of the facilities were better at hand for the disposal of the shop personnel. The general inventory of flowers in the White House flower shop was one far superior to that of the grounds. Upwards of one hundred species were not uncommonly integrated at one time or another into expert preparation for the First Family and for the dozens of rooms within the mansion. Head florist, Elmer Young (Rusty), had in fact devoted parts of his college training to the study of flower arrangement and design. His assistant, Jimmy Nelson, was also

Azalea

Lily

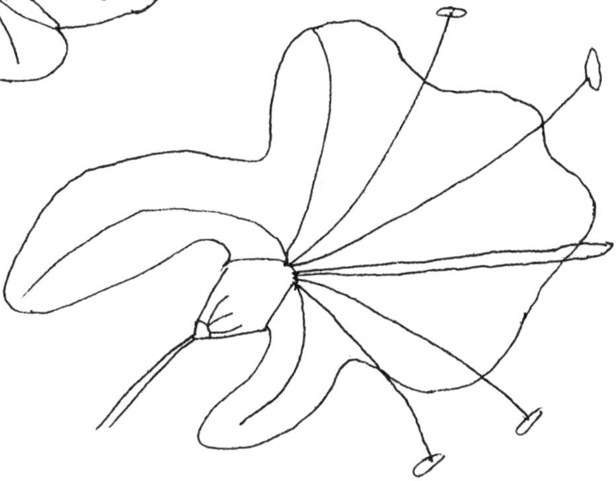

extremely adept with floral creativity and Carl Beam, a 1968 White House arrival from the Capital Parks, exhibited fine apprenticeship in the art.

When time permitted, I would wander into the flower shop and look over what was being prepared for the upstairs rooms. I watched as the artists would exercise their expertise with handling flowers, snipping just so much from one stem, discarding other flowers which were not quite appropriate for one arrangement, and scurrying back and forth like industrious ants trying to produce a propitious showing for the First Family. The Mellon vermeil flower vases and pots were among the treasured *objets d' art* to be seen and used by the personnel. The vermeil was frequently used for floral decoration, and I can remember how showy the red and white carnations would be in contrast to the gold wash of the vermeil. It was natural then for me to take pictures of the various flower arrangements created by the flower shop workers.

The Christmas season was a frenzy in the flower shop, and the normal load of personnel was expanded to include the grounds staff as well as people selected from in the house. Christmas chores were fast and many, as most all of the rooms within the White House had to be decorated before December 24. These rooms were decorated with the small semiartificial Christmas trees referred to elsewhere. The trees found their way to mantles, table centers, and on ormolu furniture. Sphagnum being what it was for moisture retention, these trees would last well into the new year without the foliage falling off. Variegated holly was a popular decoration medium in the flower shop during the Yuletide season, being used throughout the mansion on mantles, hearths and for lovely centerpieces. The holly also was placed on chandeliers and wall sconces and was frequently mixed with spruce at fireplaces. I know how hectic this season can be, having spent scores of overtime hours decorating the first floor of the White House for Christmas. I must have spent half a night trying to decorate the massive chandelier in the State Dining Room and getting it just-so right for Mrs. Nixon's inspection. The wall sconces in this room were easier to decorate because they were smaller and did not require so much effort. The upright torches in the East Room were a bit more difficult to trim, as long branches of evergreens were needed to balance between the lighting fixtures. More often than not, a perfectly decorated torch would suddenly shed its evergreen dress and a pile of greenery would be sprawled on the floor. Torches of the Grand Cross Hall had concave bowls facing the ceiling, and, as such, could receive decoration without the evergreens slipping out.

The North Entrance chandelier, a cylinder with fine glass on sides and bottom, could be decorated by the flower shop personnel by simply dropping small parcels of evergreen into the light. The six marble Doric columns separating the entrance hall from the East/West Hall were decorated with ropes of mistletoe, which was quite effective. Dozens of poinsettias, both red and white, were featured at Christmas and were set out in all of the state rooms on the first floor. Poinsettias flanked *Ficus* trees, cut evergreen trees, and were placed in bunches at various focal points.

One of the most important aspects of decorating for any occasion is presenting a beautiful, harmonious and interesting round of displays for the First Family and other visitors. At Christmas, 1968, the Christmas Tree of the Fifty States was placed in the North Entrance Hall. This tree contained a ball for each of our fifty states plus other decorations that, when assembled, made a wondrous and unique tree. I marveled as I gazed upon this tree during a staff party. The tree was a German style with electric candles with a golden star of pinpoint cream-colored lights on its tip top with red, gold, aqua, and blue balls representing the states. This tree was also encircled with tinseled ropes that glistened as I walked by.

During the regular season, the White House flower shop dealt with many of our favorite annual plants, some of which can be discussed herein. Among the many flowers to pass through the flower shop were the *Antirrhinum majus* or upright snapdragon, which was either raised at Kenilworth or bought elsewhere from local concerns. These snapdragons came in several colors, and the "Defiance," a yellow, when used with *Anemone coronaria* or Florists' Anemone, and white carnations, produced a fine display for any executive table. Add a few cuttings of budding *Delphinium consolida*—Larkspur, and of purple color, to finish off this place setting. The hurricane-lamp special, using white and yellow daisies together with some of their green foliage, makes a simple but effective setting, as was used for several White House tables in 1970. In the above case, the hurricane glass made the furnishing look topheavy, so a girdle of the above daisy gave balance to this creation. The wine-goblet arrangement, using yellow daisies, white daisies, and red tulips, also was used for place settings in the mansion. With a vermeil vase, the flower shop would do the following: *Salvia patens*—Gentian Sage of sapphire blue used to give the arrangement projection by pointing its flowers head outward. In toward the center, the red and white carnations were utilized together with yellow dahlia—"Claire deLuni" or "Py"—Pompon. To this, Rocket White Snapdragon was added.

There was an emergency in the flower shop in 1969 wherein the supply of certain flowers was dangerously low, so Rusty resorted to plucking *Pelargonium zonale*, or *Zonal Pelargonium* (geranium), and bunching them for spur-of-the-moment place settings. Mrs. Nixon was having an important luncheon that day. Geranium blooms, if not too mature, will last through a party before losing petals. They can be used in small bowls of water or as clumps in a small vase. The flowering tobacco *(Nicotiana sanderae)* can be incorporated into a nice display with the "Peeping Tom," Daffodil *(Narcissus cyclamineus)*, and Globe Candytuff *(Iberis umbellata)*. Blue Squill, or *Scilla sibirica*, by themselves make a beautiful arrangement or are nice with blue daisy *(Felicia amelloides)*. The *Felicia* was much in demand by the flower shop, and we never seemed to get enough of it.

The interior floral decorations of the White House prepared by the flower shop would always reflect what the season of the year was and what was transpiring in the gardens outside. There was a deviation in winter, because winter never came to the indoors. The First Family wanted fresh flowers year round. Even though the outside was cold and gray, the inside would look and smell sweet as spring.

Some of the annual flowers used in the White House Flower Shop not heretofore mentioned included:

Callistephus chinensis—China Aster in red, purple, pink, and cream.

Centaurea cyanus "Caerula"—Cornflower in red, blue, and pink colors.

Chrysanthemum coronaruum—Annual mum in doubles, yellow and white.

Dianthus caryophyllus or Chabaud Carnation.

Fuchsia "Citation"—Red bottom, white top.

Fuchsia "Lord Roberts"—Red and purple.

Fuchsia "Amy Lye"—Red tops and white bottom.

Gaillardia "Picta,"—Blanket Flower.

Gypsophila elegans, Chalk Plant, Baby's Breath—in white and frequently used in flower shop productions.

Helianthus annuus "Autumn Sunshine"—Sunflower.

Helichrywum bracteatum "Monstrosum"—Strawflower.

Heliotropium peruvianum—Cherry Pie, Heliotrope.

Lantana camara—Yellow, red, etc., and also used in Capital Parks landscaping plantings.

Malthiola incana—Ten-week stock for which the floral shop had a craving.

Pelargonium peltatum—Ivy—Leaf Geranium "Mme Crousse."

Petunia hybrids:

A. *Polaris*—Purple and white.
B. *Violacea*—Full purple.
C. *White Cascade*—Full white.
D. *Cajni*—Light purple.
E. *Red Cascade*—Light red.

Phlox drummondii—Annual Phlox.

Salvia patens—Gentian Sage—Sapphire blue.

Salvia spendens—"Scarlet Sage"—Blood red.

Senicio cineraria—Dusty miller.

Tagetes patula—French marigold.

A. *"Naughty Marietta"*—Brown center, yellow border.
B. *"Lemon Alldouble"*—Full yellow.

Tagetes erecta—African marigold.

Verbena hybrida—Garden Verbena—Used in Capital parks.

A. *"Sparkle"*—Red border, white center.
B. *Hybrida*—"Snow Queen."

Viola Pansy

A. *International Prize Mixed*—yellow border, black center—Used in East Garden.
B. *"Roggli"*—Purple border, black center.

Chionodoxa luciliae—Glory of the Snow.

Dahlia

A. *"Sonja"*—Pompon.

B. *"Night Editor"*—Giant decorative.
C. *"Good Morning"*—Small cactus.
Gladiolus—large-flowered
A. *"Fancy Free."*
B. *"Peter Pears."*
C. *"Perosi."*
Ageratum houstoneanum—Floss flower—Blue full.
Coleus blumei—Coleus—Various colors, varigated.
Euphorbia heteophylla—Annual poinsettia.
Fragaria vesca semperflorens—Alpine Strawberry.
Lobularia maritima—Sweet Alyssum.
Senicio cruentus—Cineraria—Various color combinations.
Hyacinthus orientalis
A. *"Ann Mary"*—Light pink.
B. *"King of the Blues"*—Light blue/lilac.
C. *"La Victoire"*—Pink.
Lilium tigrenum—Tiger Lily.
Muscari botryoides—Grape Hyacinth—blue.
Narcissus poeticus—"Queen of Narcissi."

A short listing indeed, of annuals used by our flower shop at the White House! Following is a list of perennial flowers that were used in the flower shop:

Achillia filipendulina—Fern leaf Yarrow and rare in the Rose Garden.

Althaea rosea—Double Hollyhock.

Aquilegia—Columbina—Blue and also rare in the Rose Garden.

Chrysanthemums—Varied colors raised at Kenilworth and transferred to the White House gardens.

Chrysanthemum maximum—Shasta Daisy—White with yellow center.

Coreopsis verticillata "Grandiflora"—Tickseed—Butter yellow with black-pincushion center.

Delphinium elatum
A. "Blue Beauty"—Light blue with black center.
B. "Galahad"—White with pin cushion centers.
Dianthus caryophyllus—Carnation

"Golden Wonder"—Gold.
Digitalis purpurea—Foxglove—White and pink.
Hemerocallis flava—Daylily of yellow color.
Incarvillea delavayi—Hardy Gloxinia—Pink.
Campanula muralis (portenschlagiana)—Blue and white.

Flower Shop Ferns

Dryopteris thelypteris—Marsh Buckler Fern.
Malteuccia struthiopteris—Ostrich Fern.
Adiantum venustum—Maidenhair Fern.
Adiantum pedatum—American Maidenhair Fern.
Osmuda regalis—Royal Fern.

White House Bonsai

The climax of White House gardening and horticulture rests with the ancient art of Oriental, cultivated bonsai. I had not become familarized with bonsai culture before my return to the Executive Grounds in 1972. Upon my return, however, various aspects of White House living, including gardening, had changed. I was faced with the new responsibility of caring tenderly for the bonsai conifers belonging to the First Family. The bonsai at the White House were presented to the First Family by heads of the Japanese government. The art of dwarfing trees in a small container is hundreds of years old and its beginning is nebulous. The earliest known culture of bonsai, of Japanese extraction, dates from a parchment of the thirteenth century. In China, certain trees had traditionally been grown in pots for decoration and in gardens for generations. It is thought that the art of growing trees in small containers reached Japan after A.D. 1000. The Japanese art of growing miniature trees for small potted landscapes changed little as generations passed. Craftsmen learned the technique of training trunk and branches, pruning roots, pinching foliage and plant tips, and renewing growing media. This Oriental expertise was passed down from father to son, and until recently the art was considered one for a specialist. Within the span of one hundred years, the bonsai art swept the populace on a national scale. Amateur growers by the thousands studied and implemented the art and technique of bonsai culture. Soon, bonsai societies, exhibitions and symposia became commonplace, while ancient methods of maintenance were documented and broadly discussed. Since the Second World War, bonsai culture has taken a foremost seat in the horticultural ornamentation of Japan and has finally reached the shores of America. In the West, one can already visit bonsai nurseries, clubs, and can receive mail from bonsai advertisers.

The goal of bonsai is the creation of a natural landscape with the bonsai functioning as the main theme of the creation. In the composition of the miniature garden with the shrub, one must try to achieve proportion in his creation. The roots, branches and leaves must be harmonious with the entire setting. The trunk of the bonsai is the most important element of any arrangement in that one trunk too long, too short, or too bent in a certain direction in relation to the mass will destroy the natural setting strived for. However, if a trunk is gnarled, cancered, split or otherwise damaged by attrition, then the character and value of the bonsai will be perpetuated for that specimen and will add to the overall composition.

The bonsai arrangement must always look natural, and the bonsai that is tenderly brought up will show its loving care to all who look at it. Bonsai should have full sunlight and open air circulation, and after the plant has become established and more mature, moisture should be reduced.

When one thinks of bonsai, a conifer miniature tree is expected for the minature landscape scheme. Nevertheless, deciduous trees will also make good bonsai subjects because of their color change throughout the year. A natural-shaped coniferous bonsai will suggest an eternal struggle with the elements. If root exposure is desired as some bonsai show, it would give an impression of the rustic wilderness and at the same time feature a sense of plant stability. Small rocks added to such a scheme will further enhance the natural look of the minature tree.

Trees for bonsai are plenteous, and nurseries have many types that are suitable for container growing. Those found in nurseries are quite often potbound; they are already adjusted to the indigenous climatic conditions and most important of all, such potbound trees are already arrested in their growing habits. Therefore, controlling the shape of a tree will be made easier.

Bonsai trees grow more beautifully and mellow with age, and it is thereby suggested that a long-living tree be selected, one that can adjust to the process of dwarfing and has a rugged temperament. The pine is by far one of the best trees for bonsai culture because of its being insistent on survival. Pines are seen in some of the most peculiar growing places, out of rock joints, on top of jagged cliffs, on beachheads, resisting strong and vicious winds, and growing out of sand dunes. Pines also subsist on poor soil and in temperature extremes. The pines, then, are naturals for the minature-tree-growing business. The varied species with short needles are most adaptable to miniaturing effects. The *Pinus parviflora,* or Japanese white pine, of which we had two at the White House, each approximately seventy-five years old, is one of the best choices in the pine group. The White House collection also boasted one Mugho pine *(Pinus mugo mughus),* which is a popular type for bonsai. Pines for bonsai have needles that are small and are in proportion to the dwarfed tree.

Bonsai can also be created from other types of coniferous evergreens. Among these are the juniper, spruce, and fir, all likely subjects. Many of Japan's oldest and most regal bonsai are the Sargent's juniper *(Juniperus chinensis sargenti).*

The ume or Japanese apricot *(Prunus mume)* is another popular bonsai. The ume is trained quite easily, and its redolent blossoms are a perfect addition to the dwarfed tree.

Small-leafed deciduous Japanese maples are also popular as miniature trees as are the hawthorn, quince, Japanese elm, beech, ginko and Japanese birch. Other plants such as azaleas and camellias lend themselves well to bonsai treatment. The Boston Ivy *(Parthenocissus tricuspidata)* and small leaf Boston Ivy *(P.t. veitchii)* can be trained into interesting forms.

The Japanese have four sizes of bonsai trees. These include:

Mame—Bonsai which are never over six inches in height and which have been popular for over a century. The Mame-bonsai (*mame* means "baby") are hard to raise and are attempted by relatively few people.

Ko-Bonsai—These bonsai are from six to twelve inches and are perhaps the best loved of all the bonsai. These bonsai must be in proportion to their containers as all bonsai must, a suggested ratio of one to three is well to keep in mind.

Chiu-Bonsai—This type grows up to two feet high and is an easy size to train. Containers for this group are varied in availability for the connoisseur.

Dai-Bonsai—These bonsai are the largest of what is cultivated and are over two feet high.

Bonsai pot selection is important, and it is a good idea to buy a container that does not have colored art that will distract from the simplicity and natural look of the miniature bonsai garden for which the container will be used. The White House always used simple brown unglazed and green glazed containers free of decoration. Our containers had several drain holes that permitted excess water to run out. A suitable method of covering drain holes to prevent soil washing out is to cover the holes with scraps of broken flower pots or a stone or a piece of fine mesh screen. Upon completion of this, add a layer of gravel to the bottom of the bonsai pot which will also help to retain soil during watering. A fine layer of sand is optional over and above the gravel. A layer of specially prepared soil is added next. In the case of the coniferous evergreen bonsai, a 50-50 mixture of sandy loam and sand containing moderate amounts of peat moss should be added to the pot. Use fresh soil and not that of the original pot, since the new bonsai will be growing in a minimum of space, and the medium must contain nourishment in order for the bonsai to begin its new growing pattern correctly. The above soil mixture is important, since two conditions will have to be met: the retention of soil moisture for a small area of growing and the flow of soil air.

If deciduous flowering and broadleaf bonsai are to be used, then a slight change in soil is recommended. These trees prefer less sand and more organic materials—up to 25 percent by volume.

When removing the plant from its old container, do so in the shade, and root prune all of the old and dead roots, being careful not to damage the young root hairs. Also remove as much of the original soil from around the plant as possible. Upon completing the above procedures, place the plant in the desired position in the container, and begin adding soil around the plant. To insure that dirt reaches all parts of the plant roots, a pencil or small stick can be used to lightly press soil down among the roots. There is no harm if some of the roots of the newly planted tree show above soil line. Such roots will help the tree to adjust to the new environment and will add individuality to the future bonsai plant. The new transplant should not, however, be planted too deeply as it may smother in the process.

After completing the transplant, use a fine mist of water to drench the pot until water runs out of the drainage slots. Use untreated water. Protect the new creation from sunlight for at least one week so that tree roots can once again become active and readjusted. During this time, keep the roots moist with frequent misting. Gradually place the plant back in the elements, a hardening off, so that it will be fit to be exposed.

The bonsai is a typical outdoor plant and can occasionally be relegated to the indoors for a "show and tell" session with friends. The bonsai should be kept in an environment closely approximating that of the tree before it was transplanted. Brutal sun and strong drying winds will damage the plant, as I found out when one of our Ko-Bonsai died in 1973. Fortunately, the tree was only ten or fifteen years old and not much concern was spent over it. The White House bonsai were kept in a small nursery near the east end of the East Garden which was specially prepared and coveted by George H. Hundsaker. From here the trees were taken to various garden tabletop perches, one of which was on the patio of the Rose Garden. It was in this garden that I took a few Brownie 126 photographic prints of the bonsai—one snap of a bonsai sitting on the garden table in the Rose Garden patio and another when I had gardener Dale Haney take a picture of me holding a seventy-five-year-old bonsai on the Rose Garden lawn. The latter bonsai was one of our treasured Japanese pine bonsai, and I had heard that it was worth somewhere in the neighborhood of two thousand dollars.

When shaping a bonsai, wire can be used in the process to hunch the tree or make the tree look smaller. Branches can also be wired to bring them into better balance in relation to the whole. Wiring also enables one to change the tree's direction or any portion of the tree. Since our bonsai at the White House were Ko-Bonsai, we did not wire them, but they did undergo shaping by pruning and pinching. Pruning helps the trees to maintain their rustic shape and should be done in the spring. Clean, even cuts must be made when pruning so that the design is not spoiled. Evergreens, including pines, will be limited in their growth and will propagate more branches when the newer green tips on each branch are removed by pinching back. A few evergreens continue to raise branches and new shoots during the growing season, and these should be removed. Sometimes, however, these new shoots can foster a change in the direction of the tree and can thereby be left to continue growing, if so desired by the owner.

Bonsai are typically year-round outdoor plants, and winter chill can be detrimental to them unless precautionary measures are taken to shield them from extreme cold. During the winter months, the bonsai should be watered in the morning so that this moisture can be used by the plant before dark. The bonsai must also be heeled in during severe cold, which is to say, place the bonsai and its container in a bed of peat moss or in hay or a bale of straw. This will prevent the little tree and container from freezing and breakage. Furthermore, it is suggested that the bonsai be placed on the south side of the house, where freezing during the nocturnal hours may be kept to a minimum.

If the bonsai is receiving reasonable care, then there is less chance of its being overcome by disease and insects. If insects appear, nevertheless, then it would be advisable to try a tweezer picking the insects from the tree instead of using a harsh spray.

Fertilizing the bonsai should be simple and never overdone. The theme of bonsai culture is to keep the miniature tree in its dwarfed size. Therefore, small doses of liquid fertilizer in spring and two well-spaced fertilizer applications in early summer will suffice, and the bonsai will never gain too much superfluous growth.

Bonsai

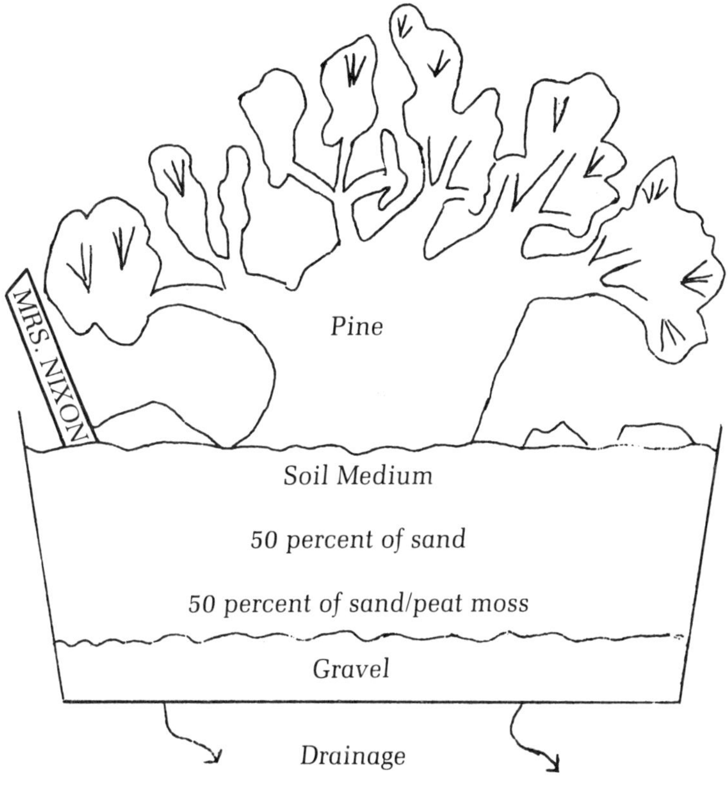

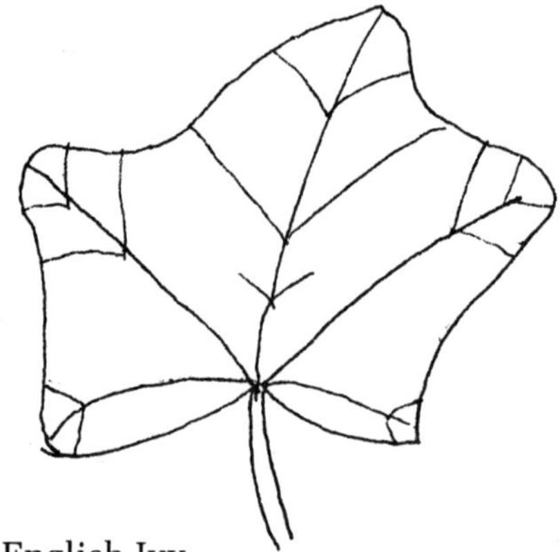

English Ivy

Japanese spurge

Purple winter creeper

Ginko biloba

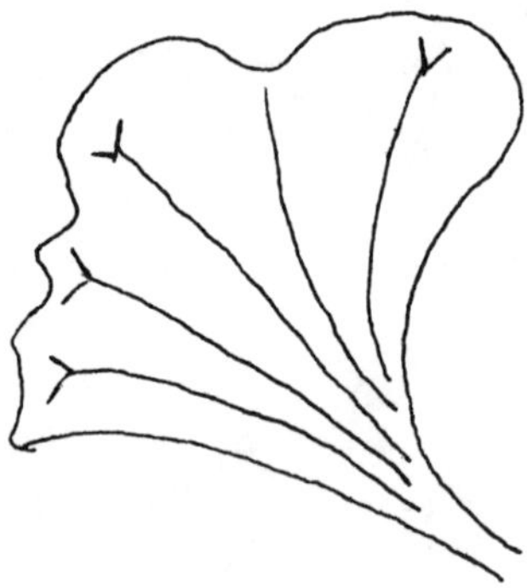

Common Defects in Bonsai Forming

Overleap

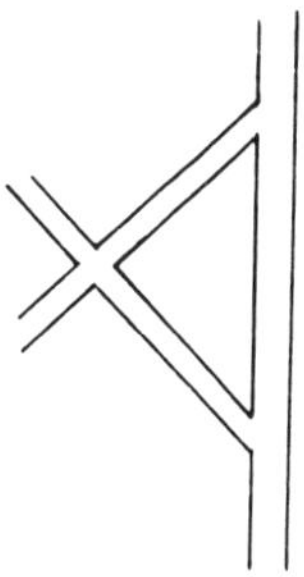

Canker

Sinuous

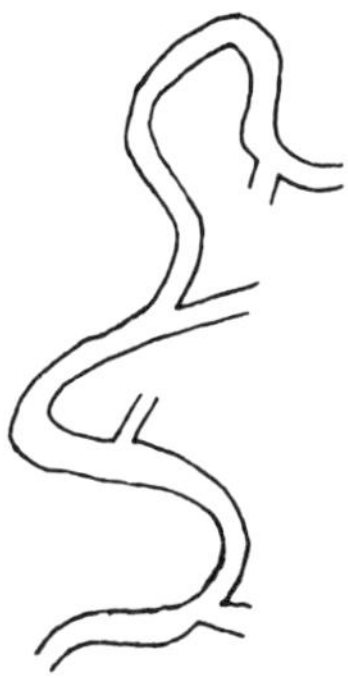

Crooked

Rotary

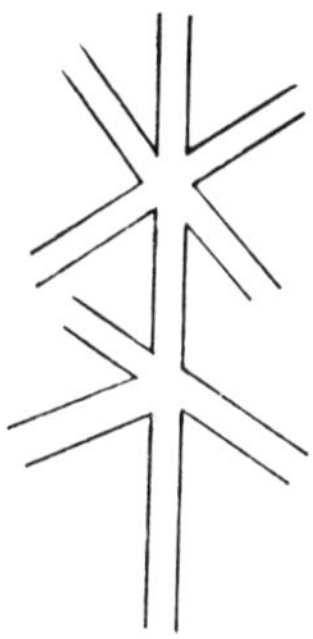

Dove Bosom

Presidential Trees

In this treatise of my White House horticultural memoirs, it is by all means apropos to include some information on the many overstory trees that have been planted over the decades by some of our country's presidents. I have at one time or another paid special attention to each of these most famous trees on the eighteen acres. Many of the White House trees have been planted, lived, and died over the years; over a dozen trees were lost during my time here. Among those that I hated to see pass was one very large and ancient American elm that had a trunk diameter of over nine feet and which was planted about the time of the burning of the White House by the surge of the British through Washington in the summer of 1814. This majestic tree graced the lower South Lawn. However, extensive digging for underground cables during the early 1970s and subsequent piling of heavy dirt around the base of the tree caused it to die by autumn of the same year—a trauma I am sure. This was a tree well over 160 years old, which, when cut, took only three days. I watched as the crown of this mighty tree came crashing to the ground under the steel teeth of the treeman's saw.

The next oldest tree is the American elm planted by John Quincy Adams during the mid 1820s, which shades the East Mound laid out by Thomas Jefferson and Benjamin Latrobe before 1820. It is amazing that this tree has survived so many years on the South Grounds, since the American elm suffers drastic consequences from the Dutch Elm beetle. The Capital Parks tree crew had been responsible for the maintenance of the tree, and over the years this care has proven beneficial in stemming whatever disease might attack the tree.

In 1829, the Magnolia grandiflora, which proudly stands by the west side of the South Portico, was planted by Andrew Jackson. This tree is remarkable in that southern magnolias may not live to a ripe old age as far north as Washington, D.C., and that this tree has seen much pruning, wiring, and carie filling of branches and trunk in the twentieth century. I also witnessed the pruning of this tree and had asked that portions of the wood be saved. After so many years of growing on the White House grounds, the Jackson magnolia has seen fit to produce beautiful flowers and to withstand the flocks of squealing starlings.

During the administration of Benjamin Harrison—March 4, 1889 to March 3, 1893—the North Grounds of the White House received a Presidential gift—a beautiful Scarlet oak—which today is over sixty feet high and stands in the northeast panel. This tree is one of the more regal specimens of the hard woods to shade the lawn and has had many years of vigorous growth.

The Woodrow Wilson American elm which was planted during his term in office at the White House—March 4, 1913 through March 3, 1921—graciously enframes the east end of the mansion when seen from Pennsylvania Avenue. This elm is one of the healthiest trees on the North Grounds and has responded well to root fertilization.

During the mid 1960s, seeds were taken from the Warren G. Harding southern magnolia and delivered to Kenilworth for propagation. By the end of my second year at the Kenilworth facility, I had already gotten my Warren G. Harding magnolia seedling of two feet in height and planted it at my home in Maryland. The W.G.H. magnolia, together with another specimen of the same type, raise their shiny leaves and set off the tourist entrance on East Executive Avenue. This magnolia is over fifty years old and was planted during the Harding Administration—March 4, 1921, until August 2, 1923.

Other Presidential trees planted during the various Presidential terms of office include:

Japanese Maples—Planted on the Lower South Grounds, and which bring viewers' attention to the large South Fountain. The maples were planted during the Grover Cleveland term—March 4, 1893 to March 3, 1897.

European White Birch—An addition to the Lower South Grounds during the Calvin Coolidge years—August 3, 1923, until March 3, 1929.

White Oak—Situated in the west center panel of the South

Grounds, was a gift to the White House from Herbert Hoover during his four-year term of office—March 4, 1929 to March 3, 1933.

Little Leaf Lindens—Which are growing in the upper center panel near the South Portico, were gifts during the Franklin D. Roosevelt years—March 4, 1933 through April 12, 1945.

Northern Red Oak—Growing to the left of the South Portico, as one looks south, and planted during the Presidency of Dwight D. Eisenhower—January 20, 1953 to January 20, 1961.

Apple—Standing near the Eisenhower Northern Red Oak, and a gift from the Kennedy Administration—January 20, 1961, until his untimely death on November 22, 1963.

General tree care a la White House was usually the responsibility of the NCP tree maintenance crew. This care included tree-spraying, pruning, tree removal and in the latter 1960s for the large American elms, crotch drainage. The latter was a vital function whereby a copper drain pipe was inserted into the elms just below the crotching to relieve accumulated water from these spots. Tree crotches tended to fill with water over a period of time from rain and snow-melt, to a point where the crotches would overflow gradually and quietly and leave an ugly stain running down the trunk of the tree. Our method of drainage was to simply drill an air hole into the crotch from above, thus admitting air into the watered cavity. From below the tree crotch, a second hole was drilled and the water would flow free of the crotch with air behind it. Next, a half inch pipe was inserted to act as a permanent drain for future seepage. The original air hole was plugged with a wooden peg to prevent aeration and rot of unprotected wood. Finally, plugs, pipe, and any other blemishes were treated with pine tar to make the trunk look more natural. The above crotch-drainage method can and was multiplied according to the size of the tree trunk and volume of the crotch cavity to be emptied. One of our elms had three drainage pipes protruding from the trunk and several black splotches where air vents were plugged.

Tree Crotch Drainage

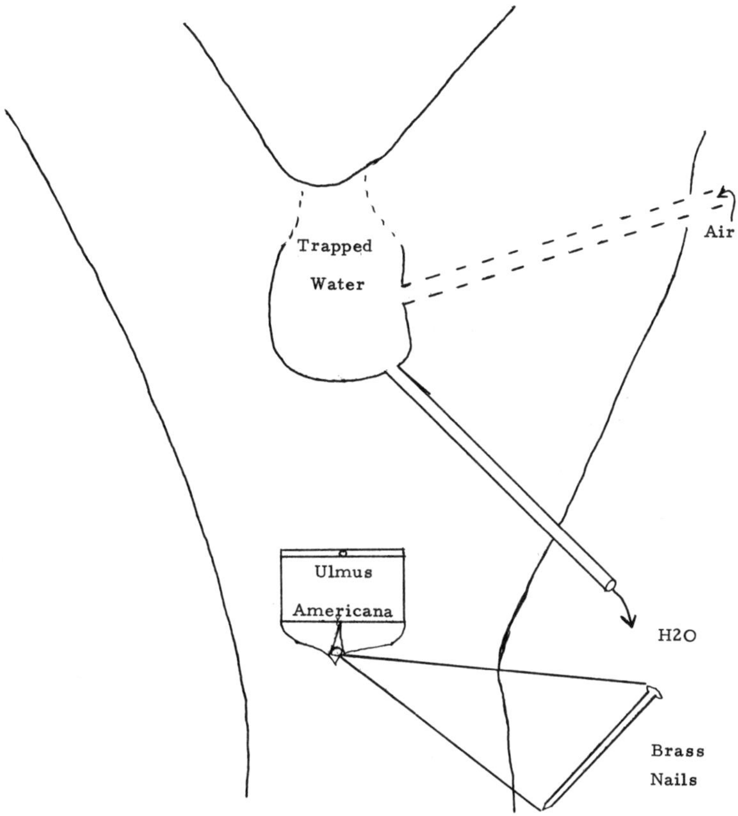

Tree fertilizing at the White House was one of our projects and was done during the winter. The larger trees required up to three hundred punched holes in a spiral network beginning ten feet from the trunk and moving outward to the farthest branches. Holes were approximately fifteen inches deep, and in each hole was placed a trowel full of chemical fertilizer; the 10-6-4 formula was used. The hole was then filled with a trowel or two of common soil and thereafter stepped on to make the ground firm. Fertilizer was applied according to inch diameter of tree trunks. I spent countless hours with the large American elms on the South Grounds and also with the elms and oaks on the North Grounds. Punching eighteen inch holes into the ground with heavy iron bars was tiresome, and I boasted more than one blister on my hands. Gloves, à la G.I., were not gloves at all, but cheap, brown, cloth covers that would fray and become full of holes with the least tension. I went through four pairs of gloves in one winter and was about to snitch a fifth when I decided to buy my own pair of heavy-duty gloves to finish the arduous tree feeding task.

The White House gardening exposé has come full cycle and has dealt with interesting facets of horticulture as seen by my staff and me. It was an honor and joy to have been selected to work on the White House Staff and to have partaken of its horticultural diet.

Magnolia grandiflora

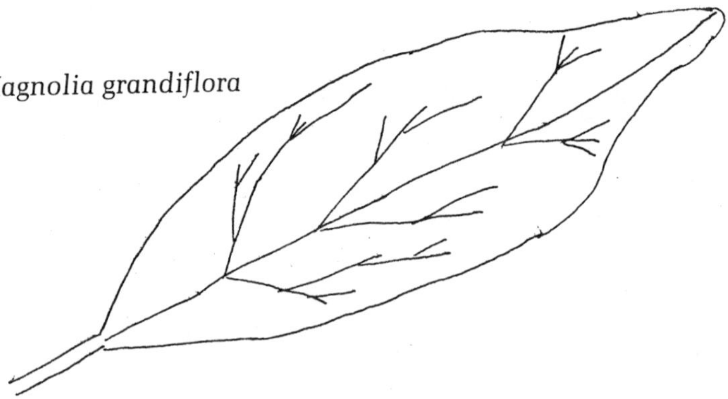

Garden Schemes

The North Fountain

Flower arrangement, 1968-1969

I. Autumn of 1969
 A. Mums
II. Winter—Nixon Administration 1969–70
 A. Spreading yews in tin cans—middle bed
 B. Silver gray Saint Helenia—inner/outer borders
III. Spring—Nixon Administration 1970
 A. Potted tulips—Kenilworth
 B. Yews pulled, boxed hyacinths planted as border
IV. Summer—Nixon Administration 1970
 A. Potted hyacinths, potted tulips pulled
 B. Geraniums planted—middle bed
 C. Dusty miller—inner/outer borders

I. Autumn of 1968
 A. Mums
II. Winter—Johnson Administration/Nixon Administration
 A. Red tulips—planted naturally
 B. Blue hyacinths—trenched, natural plant
III. Spring—Nixon Administration 1969
 A. Tulips/Hyacinths—natural bloom.
IV. Summer—Nixon Administration 1969

A. Hyacinths pulled
B. Tulips pulled
C. Geraniums planted
D. Dusty miller planted—outer/inner borders

Floral Display, 1968–69

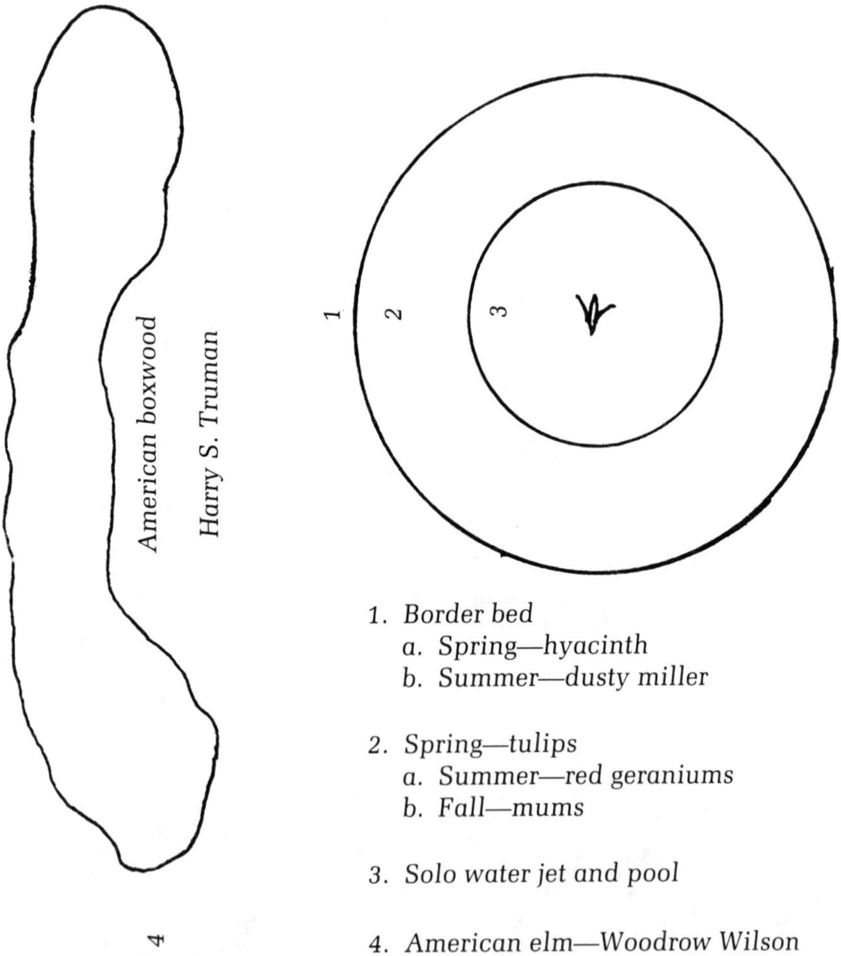

1. Border bed
   a. *Spring—hyacinth*
   b. *Summer—dusty miller*

2. *Spring—tulips*
   a. *Summer—red geraniums*
   b. *Fall—mums*

3. *Solo water jet and pool*

4. *American elm—Woodrow Wilson*

The South Fountain

Floral Display, 1968–1969

I. Autumn of 1969
   A. Mums
II. Winter—Nixon Administration, 1969-1970
   A. Canned yews planted
   B. Saint Helenia—inner border, planted
III. Spring of 1970
   A. Potted tulips planted
   B. Yews pulled, Saint Helenia pulled
   C. Pansies pulled
IV. Summer of 1970
   A. Tulips pulled
   B. Geraniums or marigolds planted
   C. Dusty miller or petunias—inner/outer edge

I. Autumn of 1968
   A. Mums
II. Winter—Johnson/Nixon Administrations, 1968-1969
   A. Tulips planted
   B. Pansies—borders
III. Spring—Nixon Administration, 1969
   A. Red or yellow tulips
   B. Pansies
IV. Summer—Nixon Administration, 1969

A. Tulips pulled
B. Pansies pulled
C. Marigolds or geraniums planted
D. Dusty miller—inner/outer borders

South Fountain Floral Display, 1972–1973

I. Autumn of 1972
  A. Marigolds pulled
  B. Mums planted
II. Winter of 1972
  A. Tulips planted
  B. Pansies—inner/outer borders
  C. Mums pulled, dusty miller pulled
III. Spring of 1973
  A. Tulips bloom
  B. Pansies bloom
IV. Summer of 1973
  A. Marigolds—middle border
  B. Dusty miller—outer/inner borders

North Fountain Floral Arrangement 1972–1973

I. Autumn of 1972
  A. Geraniums pulled
  B. Mums planted
II. Winter of 1972
  A. Mums pulled, dusty miller pulled
  B. Tulips planted
  C. Hyacinths planted—outer border
III. Spring of 1973
  A. Tulips bloom
  B. Hyacinths bloom
IV. Summer of 1973
  A. Geraniums planted—mid border
  B. Dusty miller planted—outer/inner border
  C. Tulips/hyacinths pulled

South Fountain Floral Display, 1968–1969

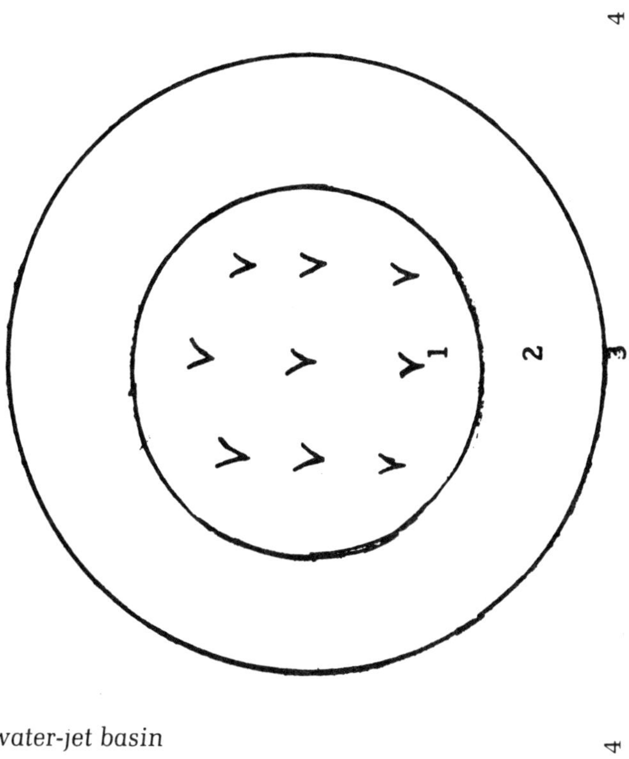

1. *Nine water-jet basin*

2. *Spring—tulips*
   Summer—geraniums
   Fall—mums

3. *Border bed*
   a. Spring—pansies
   b. Summer—petunias

4. *Japanese maples—Grover Cleveland*

The East Garden

The Pool

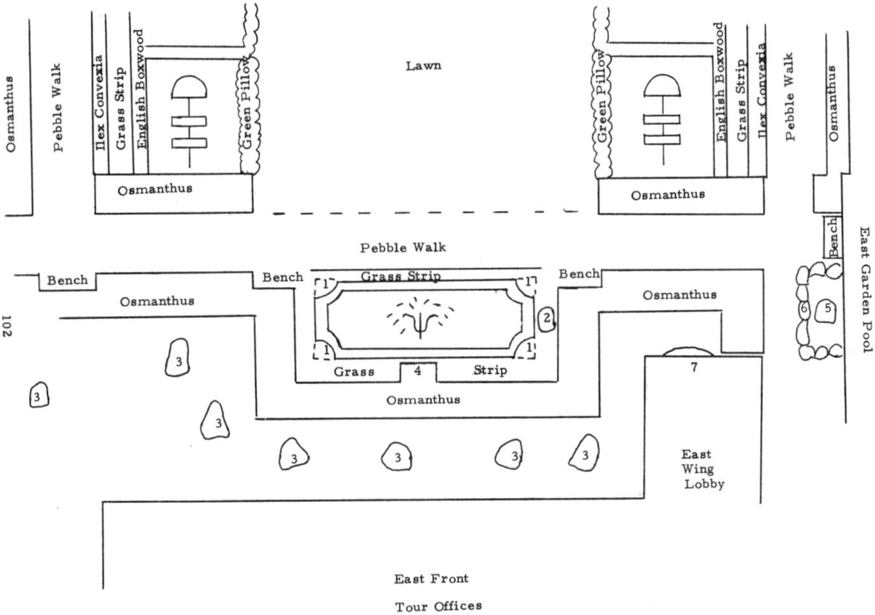

The Plants

1. Potted Herbs, annuals
    A. Spring—tulips, marjoram
    B. Summer—petunias, marjoram
2. Mock orange
3. Evergreens, deciduous
4. Water nymph (bronze)
5. Magnolia soulangeana
6. English Boxwood
7. Climbing Russian Olive

The Patio

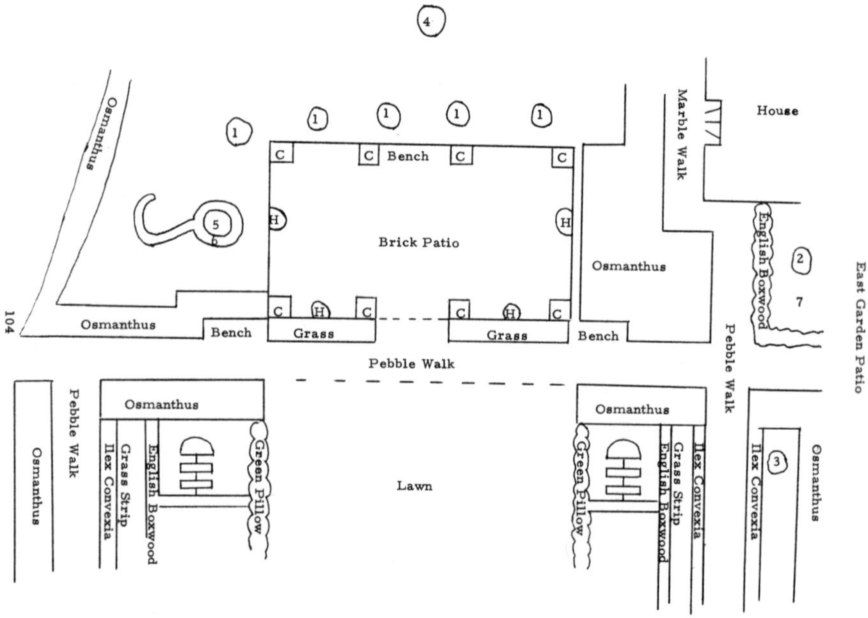

The Patio Plants

1. Evergreens
2. Magnolia soulangeana
3. Little leaf linden
4. Magnolia grandiflora
5. Hawthorn
6. Bulb bed—Tulips, hyacinths—Spring, 1973
   Begonias—Summer, 1973
7. Euronymus cover

C. Columns supporting trellis and grape plant
H. Potted herbs—annuals—spring through autumn, 1972–1973

The Floral Display

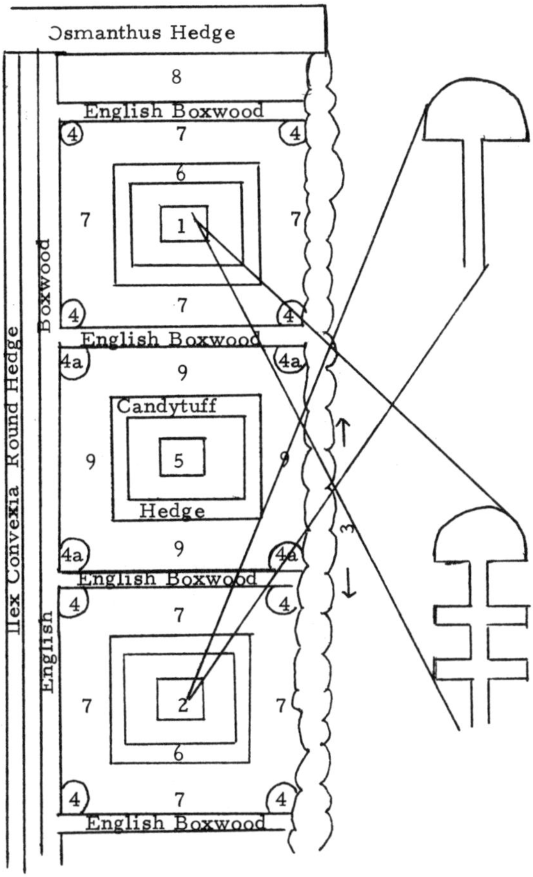

1. Topiary holly
2. Topiary holly
3. Green pillow boxwood
4. Rosemary herb

4a. Saint Helenia herb

5. Daisies—Summer
   Tulips—Spring
6. Dusty Miller—summer
   Saint Helenia—winter
7. Pansies—spring
   Geraniums—summer
   Mums—fall
8. Tulips, Allium—spring
   Marigolds—summer
   Mums—fall
9. Tulips, pansies—spring
   Geraniums—summer
   Mums—fall

Portions of the Autumn Scheme

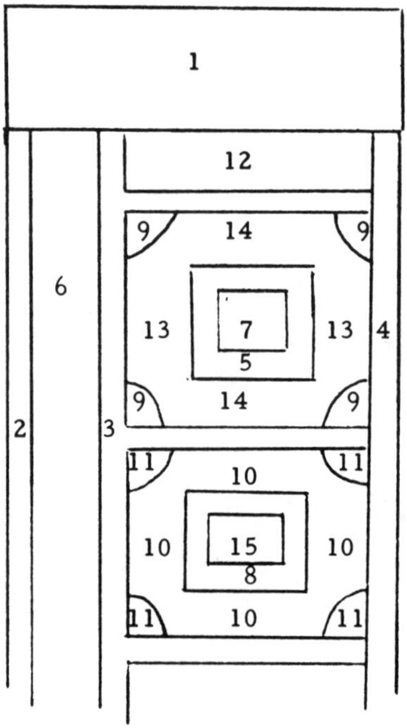

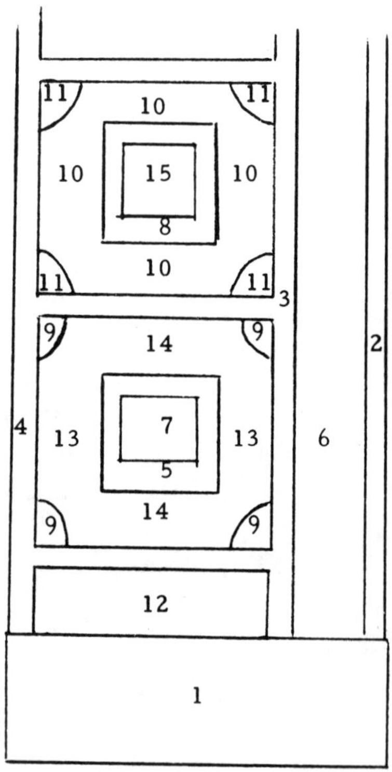

1. Osmanthus hedge (beveled)
2. Ilex convexa (holly)
3. English Boxwood
4. Green pillow boxwood
5. Thyme hedge (square trim)
6. K31 grass strip
7. Topiary holly
8. Candytuff (square unhedged)
9. Rosemary
10. Chives
11. St. Helenia
12. Mum—Yellow Supreme
13. Mum—Bronze Song
14. Mum—Pumpkin
15. Mum—Red Mischief

The Rose Garden

The Rose Garden—East

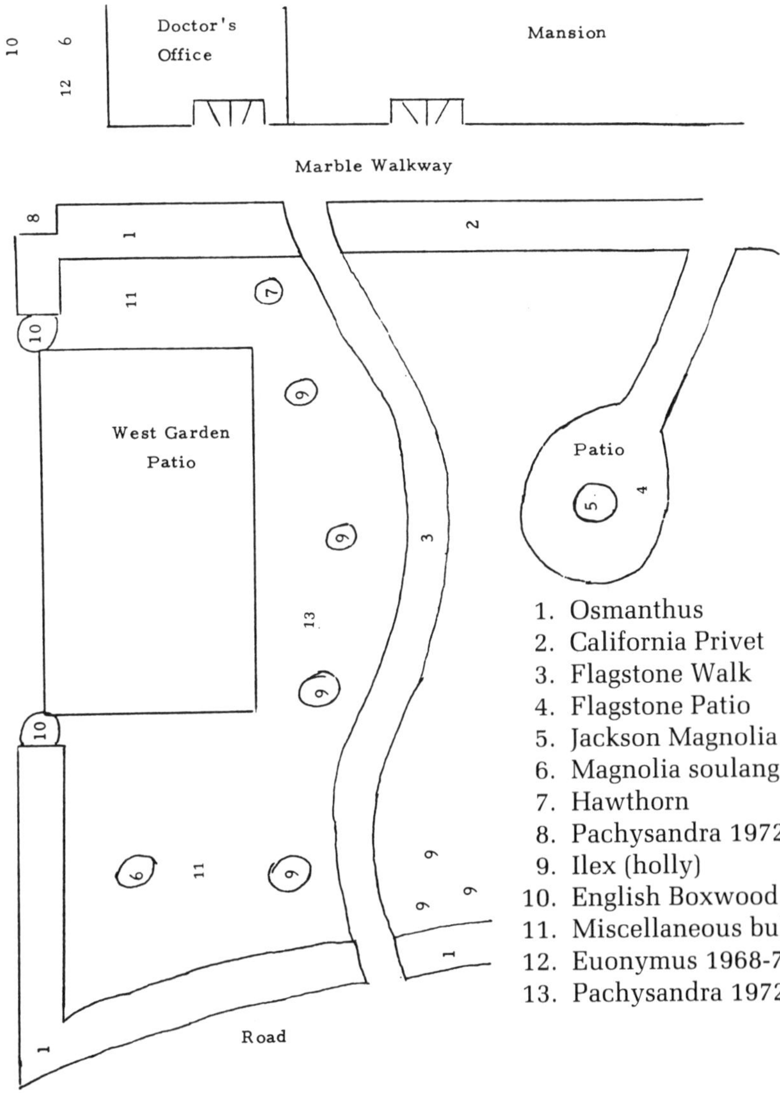

1. Osmanthus
2. California Privet
3. Flagstone Walk
4. Flagstone Patio
5. Jackson Magnolia 1829
6. Magnolia soulangeana
7. Hawthorn
8. Pachysandra 1972-1973
9. Ilex (holly)
10. English Boxwood
11. Miscellaneous bulbs
12. Euonymus 1968-72
13. Pachysandra 1972 on

The Rose Garden—West

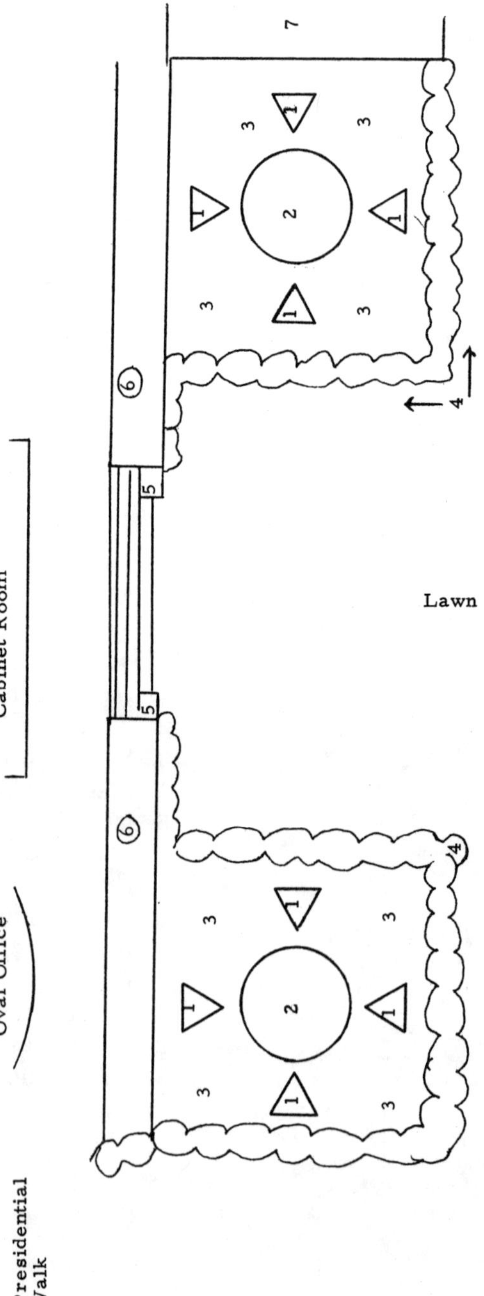

1. Saint Helenia—Winter
   Dusty Miller—summer 1972-73
2. Magnolia soulangeana
3. Begonia Rex—Summer
   Grape hyacinths—Spring 1972
4. English Boxwood
5. Poodle Juniper—1972-1973
6. Ilex (holly)
7. Osmanthus

The West Garden

Spring Bulb Scheme

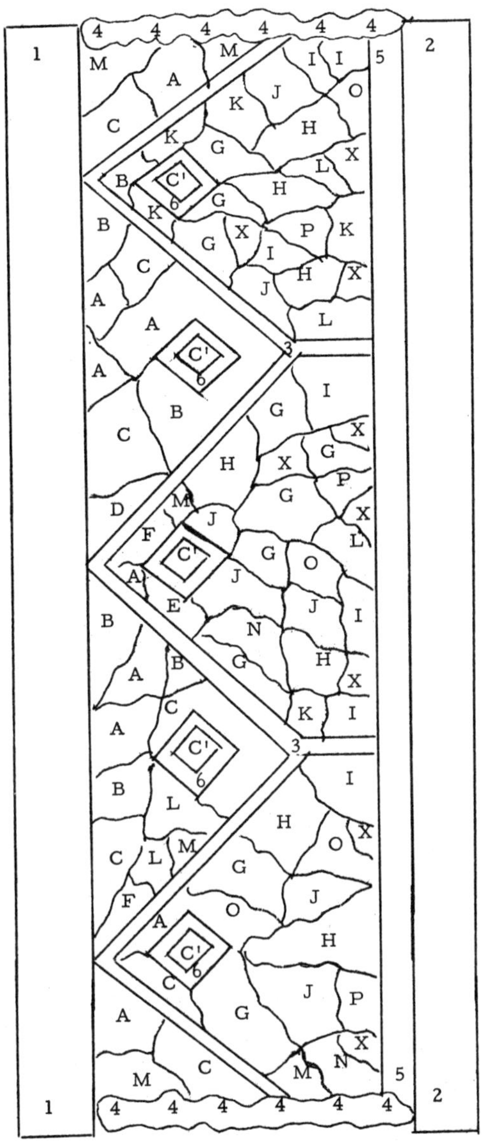

1. Osmanthus
2. Ilex (holly)
3. English Boxwood
4. English Boxwood
5. Grape hyacinth
6. Saint Helenia

A. Empire State
B. Gudoshnik
C. Lefeber's Favorite
D. Spring Song
E. Jewel of Spring
F. President Kennedy
G. Mrs. John T. Scheepers
H. Mother's Day
I. Red Emperor
J. First Lady
K. Bing Crosby
L. Ibis
M. Allium
N. Blizzard

Perennials

O. Gaillardia
P. Columbine (Aquilegia)
C¹. Flowering crab apple
X. Rose

The Annual Scheme

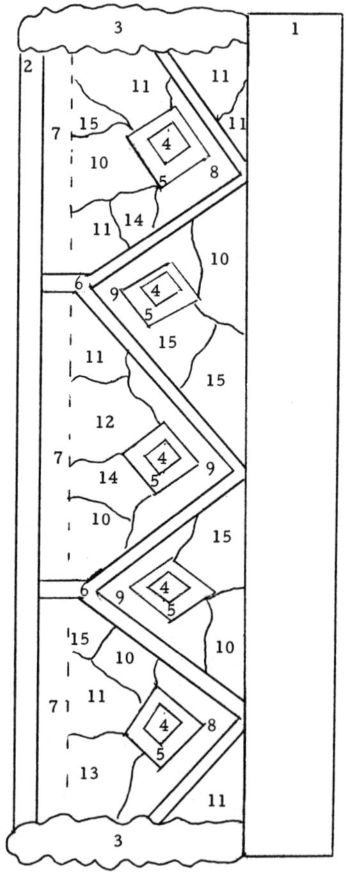

1. Osmanthus
2. English box hedge
3. English box
4. Flowering crab apple
5. Dusty Miller hedge
6. English Boxwood hedge
7. Heliotrope
8. Susan Baldwin—salmon geranium
9. Jolly Rogers—scarlet geranium
10. Improved Richard—red geranium
11. Ricard—Lorelei—pink geranium
12. Brutus—red caspium geranium
13. Paul Crampel—red geranium
14. Kleine Liebling—pink geranium
15. Irene—red geranium

The Autumn Scheme

1. Osmanthus
2. Ilex (holly)
3. English Boxwood
4. English Boxwood
5. Dusty Miller
6. Dusty Miller
7. Flowering crab apple
8. Columbine (Aquilegia)
9. Gaillardia
10. Mum—Red Mischief
11. Mum—Red Headliner
12. Mum—Indian Summer
13. Mum—Pumpkin
14. Mum—Dolli-ette
15. Mum—Bronze Song
16. Mum—Yellow Supreme
17. Mum—Jesamine Williams
18. Mum—Pink Chief
19. Roses

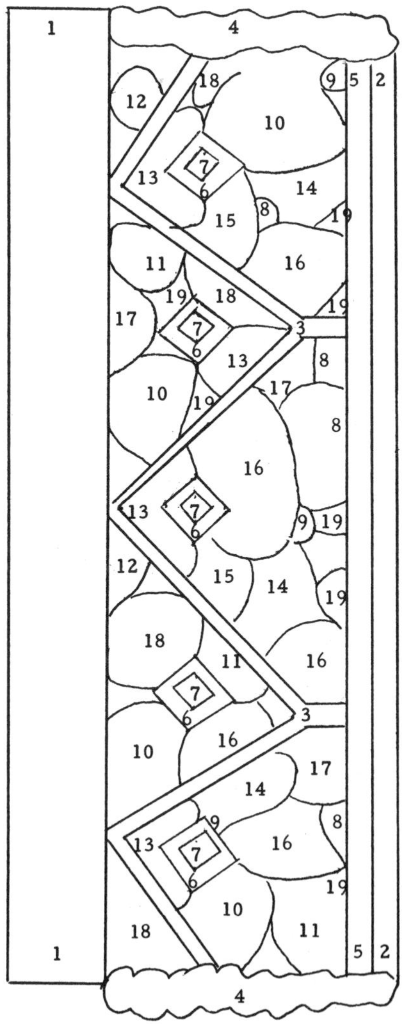

Upright Topiary Ficus

Upright Ficus (fig) often called to serve as an avenue of backup greenery during ceremonies at the South Portico of the White House.

Tubbed White Pine and Ilex

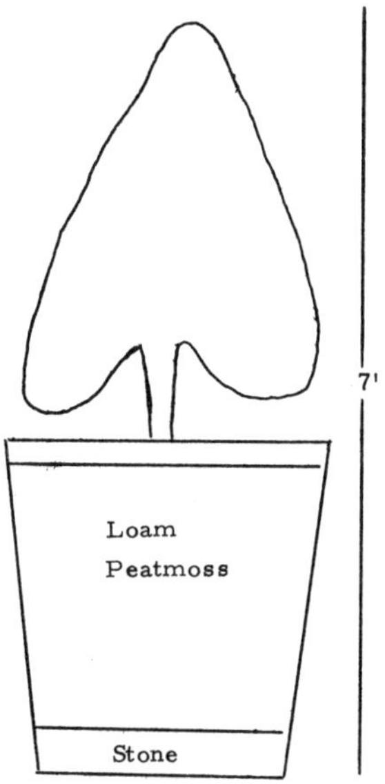

Specimens of white pine were used on the exterior side of the third floor solarium of the White House.

Dirt line at two inches below top of tub

Peat moss mulch used to increase moisture so as not to allow evergreen to dehydrate in summer heat.

Ilex convexa was also used for floral treatment on the third floor of the White House and in the Presidential Apartments.

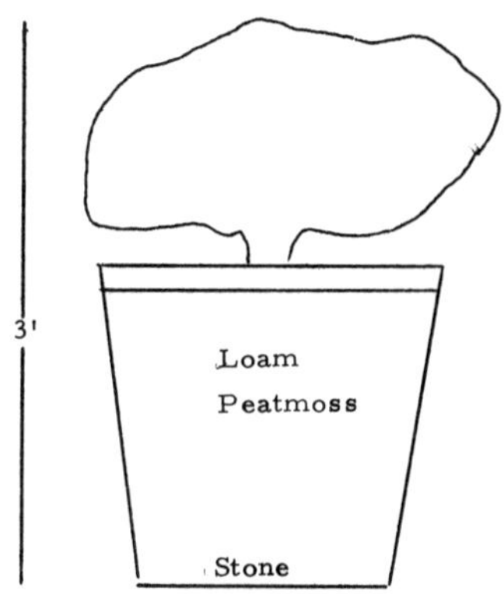

The White House Solarium

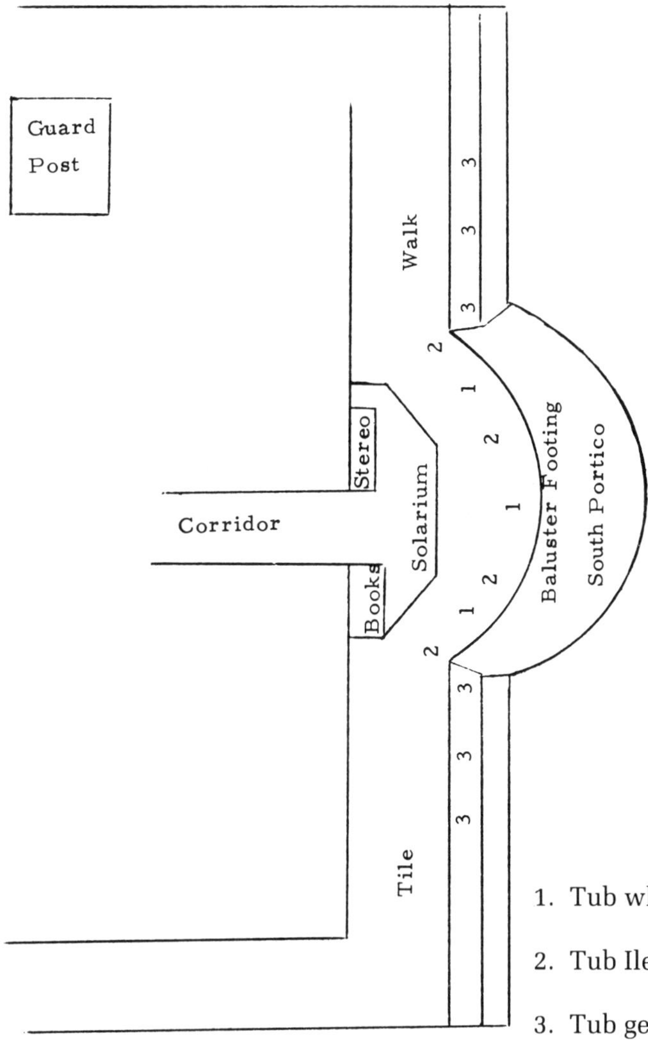

1. Tub white pines

2. Tub Ilex convexa

3. Tub geraniums, petunias—2′

Miniature White House Christmas Trees

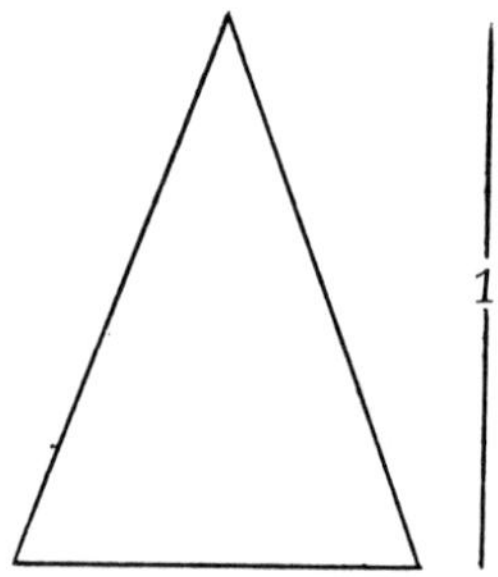

Small Taxus twigs were cut, pointed and pinioned to steel clasps and inserted into tree frame.

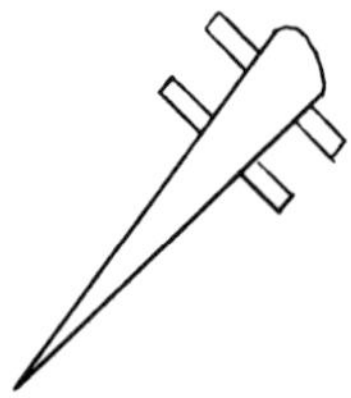

One-inch chickenwire form stuffed with sphagnum moss and soaked for twenty-four hours, drained for two hours.

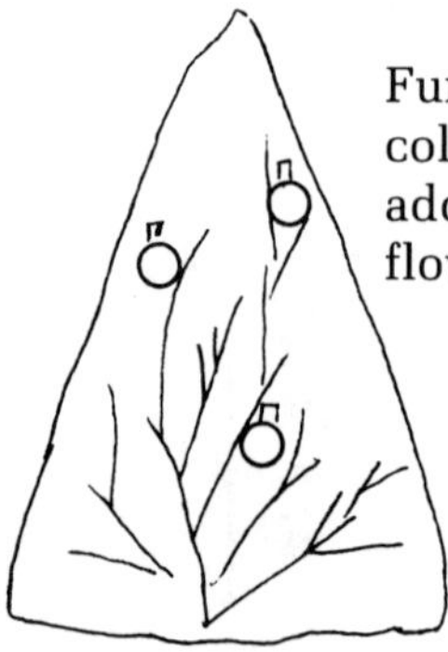

Further decorations, colored balls, ribbon, added by White House flower shop.

Finished tree with bowls of Taxus *(cuspidata)*. Final product was placed on tables in the White House rooms with an evergreen complement circling the base of each tree.

Jefferson Mounds, South Grounds

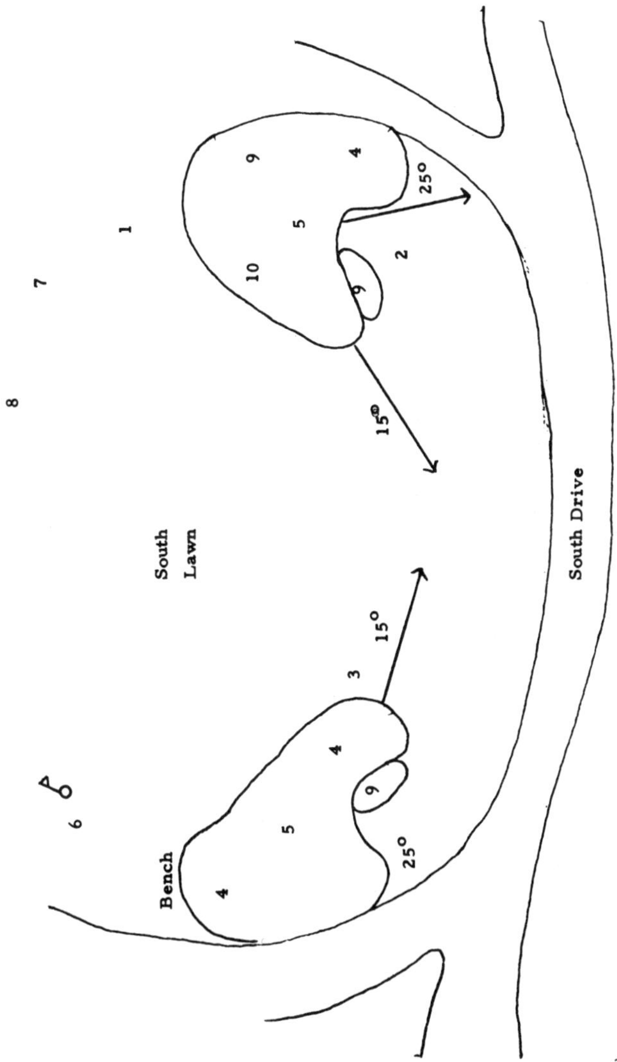

1. Horse chestnut
2. American elm
3. Pacific cedar
4. Bronze beech
5. English Ivy
6. Eisenhower putting green
7. Taxus
8. Magnolia
9. Ligustrum
10. Linden

The Children's Garden, 1968

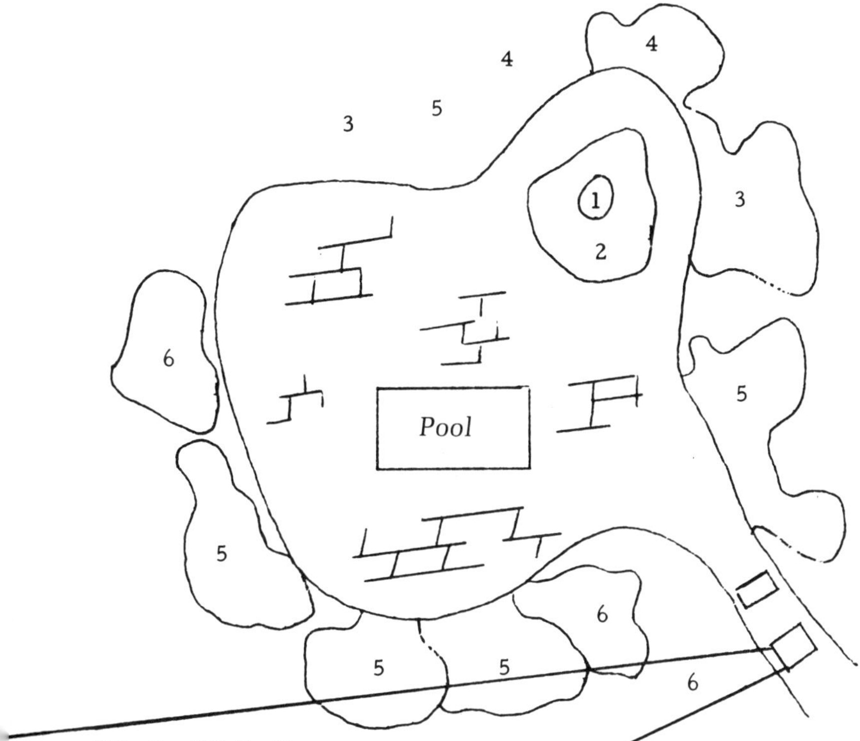

"To the White House grounds from the President and Mrs. Johnson, 1968"

1. Crab apple
2. Circular bed for bulbs and annuals
3. Ilex cornuta
4. Ilex opaca
5. Ilex rotunda
6. Ilex convexa

Glossary

Aerate (soil). To supply the soil with air by means of breaking up the soil using mechanical means.
Annual. Plants living for only one year or one season.
Apical dominance. The vigorous top growth of the stem of a plant which may increase over lateral growth.
Berry. The fleshy, indehiscent fruit which may contain more than one seed.
Bone meal. A ground animal bone used for fertilizer and containing from 20 to 25 percent phosphoric acid and 2 percent nitrogen.
Border plant. A plant species used as a border along a flower bed or along a border of trees and shrubs.
Bulb. A swollen stem which is usually located underground with fleshy leaf-bases.
Callous. The advance and subsequent covering of a plant wound by woody growth. Callousing will commence after pruning.
Coldframe. Angular box-like structure used to protect plants from cold temperatures.
Compost. A mixture of loose vegetative matter, manure, peelings and other once living plant waste that is left to decay by bacterial action. Compost is used for fertilizer and soil conditioning.
Conifer. A plant species which bears cones, *i.e.*, pines, spruces, hemlocks.
Cottonseed meal. Ground cottonseeds that are used for fertilizer. Contains from 6 to 10 percent nitrogen, 2 to 4 percent phosphorus, and 2 percent potassium.

Creeper. Generally applied to ground cover as English ivy, euonymus (purple winter creeper) and various weeds.

Crown. The top of a tree including living branches and foliage.

Cutting. A section of a plant on a root that is cut and planted in a rooting medium so that it can sprout roots and develop into a plant like that of the parent.

Deciduous. The annual shedding of leaves by a plant, i.e., oaks, maples, sycamores.

Dissectum. Dissected, divided into numerous fine, thread-like segments.

Division. Separation of roots, etc., for reproduction purposes.

Drainage. To be well-drained, passage of surplus water through a soil.

Evergreen. Persistent on year-round leaves or foliage on a plant.

Filicifolium. Fern-leaved.

Flower. The part of a plant that reproduces the species.

Full sun. A garden plot receiving full sun has no shade during at least eight hours of daylight.

Grafting. A means of propagating a plant species by slicing and inserting one plant cutting into that of another.

Half-sun. A garden receiving half sun will have shade for up to one half of a normal day.

Harden-off Prepare an indoor-grown seedling to the outdoor environment by gradually exposing it to the outdoors.

Herb. A seed-producing annual, biennial or perennial plant that does not develop persistent woody fabric, but dies back at the end of a growing season.

Hothouse. An artificially heated house of glass for growing plants during winter.

Humus. Decomposed organic matter in a soil.

Hydroponics. A system of culture in liquid nutrients.

Layering. Propagation method. Natural layering commences when plant nodes are in contact with the soil.

Leaching. Dissolved nutrients or salts out of a soil or fertilizer by the action of water filtering downward.

Leaf mold. A composition of rotted leaves, wood, seeds, and other litter.

Linearilobum. Narrow, long and lobed.

Loam. A soil of one third clay, one third sand and one third humus.

Manure. The dung of cattle used as a fertilizer. Fresh manure is best for composting until it has aged. Well-cured manure is suitable for direct application.

Medium. A substance regarded as a means of transmission of the propagation force, i.e., perlite, peat moss, sand and water.

Mildew. A parasitic fungi forming a growth of filaments on a host plant.

Mulch. Protective cover placed over the soil between plants to reduce erosion, reduce evaporation, maintain soil temperature and help reduce weed germination.

Nitrogen. One of the three most important plant nutrients; also phosphorus and potassium. Nitrogen is responsible for the production of leaves and stems of plants.

Node. Section or joint of a plant where stems and leaves grow and mature. Nodes also produce the rooting in new plants when inserted into the propagation bed.

Overstory. Consists of the tallest plants—oaks, elms, hickory.

Parasite. An organism living on a plant and extracting nourishment therefrom.

Peat. Prehistoric plant remains that have decayed under bog conditions (airless).

Peat pots. Tiny containers made of peat which may or may not contain a planting medium. Often used as an intermediary between the propagation bed and clay pots.

Perennial. A plant that grows year round.

Perlite. A type of propagation material which is artificially blown-up and volcanic in origin.

Phreatophyte. A deep-rooted plant obtaining its water from the water table or water above the water table. In horticulture an undesirable plant like the common mulberry or wild cherry.

Pinching. Snipping off in order to produce a shape of a plant or to increase flowering or fruiting. Pinching can be done with index finger and thumb.

Poodle. A form of pruning whereby the plant takes the form of having rounded concentrations of growth stripped stems in between.

Potbound. Refers to the matted, congested and spiraling of plant roots in a pot due to the lack of repotting the plant. Excessive roots can be pruned from the plant and the plant repotted in the next size larger.

Propagation (mist). Irrigation of plant cuttings by overhead watering to prevent wilt.

Prune. To cut branches in order to induce regular growth.

Racemes. Flower-clustering with separate flowers at equal distances along a stem. The youngest cluster is at the top of the stem.

Root. Part of a plant below ground which serves as physical support and which supplies nutrients from the soil to the plant.

Runner. A creeping stem growing from the main stem and rooting at the nodes.

Sand. Small, worn particles of silica and other rock types less than two millimeters in diameter.

Scion. A detached living portion of a plant joined to a grafting stock. A cutting.

Seedling. A young plant grown from a seed.

Septemlobum. Seven lobes.

Shrub. A woody plant that will branch near the grown surface.

Soil. Clay-acidic, chalk-calcareous, fibre, peatacid, sand-light.

Sphagnum Moss. Commonly found in bogs and which is used in hanging plant baskets and as an additive to soil.

Standard. A type of pruning whereby the crown of a tree is rounded while the stem is free of branching and foliage.

Stem. That portion of the plant that supports leaves and flowers.

Sucker. A shoot which arises just below ground or which generates itself from branch nodes.

Thinning. Removing young plants from a group so that the remainder can mature properly.

Topiary. Ornamental gardening whereby plants are trained, cut and pruned into odd or ornamental forms.

Trifoliate. A three-parted system of leaves.

Umbel. A flat-top raceme with flower stalks growing from a single point.

Understory. Those plants reaching fifteen feet or less and which grow under taller species of plants, i.e., dogwood, Japanese maple, *Poncirus trifoliate.*

Variegated. Designates a color combination of holly, ground covers and many newer plant species. Colors are shades of one color.

Winterkill. To die as a result of exposure to winter conditions such as drying winds and moisture-depleted soil.